Remembering

What 50 Years of Research with Famous Amnesia Patient H.M. Can Teach Us about Memory and How It Works

Donald G. MacKay

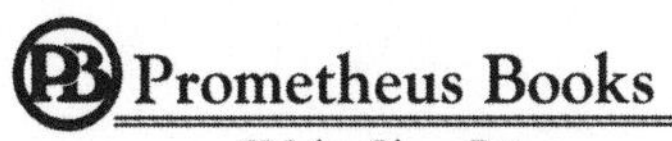
Prometheus Books
59 John Glenn Drive
Amherst, New York 14228

Published 2019 by Prometheus Books

Inquiries should be addressed to
Prometheus Books
59 John Glenn Drive
Amherst, New York 14228
VOICE: 716–691–0133 • FAX: 716–691–0137
WWW.PROMETHEUSBOOKS.COM

23 22 21 20 19 5 4 3 2 1

Library of Congress Cataloging-in-Publication Data

Names: MacKay, Donald G., author.
Title: Remembering : what 50 years of research with famous amnesia patient H.M. can teach us about memory and how it works / Donald G. MacKay.
Description: Amherst : Prometheus Books, 2019. | Includes index.
Identifiers: LCCN 2018039637 (print) | LCCN 2018054683 (ebook) | ISBN 9781633884083 (ebook) | ISBN 9781633884076 (hardback)
Subjects: LCSH: Memory. | Cognition. | Psychology. | BISAC: SELF-HELP / Personal Growth / Memory Improvement. | PSYCHOLOGY / Cognitive Psychology. | SCIENCE / Life Sciences / Neuroscience.
Classification: LCC BF371 (ebook) | LCC BF371 .M3375 2019 (print) | DDC 153.1/2—dc23
LC record available at https://lccn.loc.gov/2018039637

Printed in the United States of America

Remembering

To Henry Molaison,
an ordinary man who became famous by generously devoting his life
to helping scientists understand his memory, mind, and brain,
trusting in the promise that what they learned about him
would "help others."

CONTENTS

SECTION I: THE BROKEN MEMORY MYSTERY

SECTION II: THE ENGINES OF MEMORY AND MIND

SECTION III: AWARENESS OUT OF THE BLUE

SECTION IV: OUT OF THE BLUE CREATION

SECTION V: COMPENSATING FOR CATASTROPHE

SECTION VI: CELEBRATION AND COMMEMORATION

SECTION I
THE BROKEN MEMORY MYSTERY

Chapter 1
WE'RE ALL FORGETTERS, SAMANTHA

WHY DO WE FORGET THE NAMES OF CHERISHED FRIENDS WE HAVE KNOWN FOR MANY YEARS?

The tip-of-the-tongue experience is a classic example of forgetting. It happened to me in 2014 during an interview with a news reporter, Samantha Kimmey.[1] She wanted to interview me after reading my just-published article about Henry M., a famous amnesic patient she had heard about as an undergraduate major in psychology.[2] The purpose of her interview was to find out whether my research with Henry could help her readers understand and cope with memory failure. That simple question became the theme for this book. *Remembering* shows how normal people can maintain their memory, mind, and brain in ways that Henry could not.

Samantha interviewed me in my home two dozen miles northwest of San Francisco. Before her first question, I wanted to tell Samantha a story from thirty years earlier about the charismatic editor of her Pulitzer Prize–winning newspaper. I could hear his voice and picture his face. I just needed his name. It was on the tip of my tongue. Nothing emerged.

Struggling with my silent memory failure, I stared past Samantha out the picture windows of my office. I studied the beautiful rolling hills created by earthquakes along the San Andreas fault. Still no name.

Hiding my frustration and embarrassment, I returned my focus to the interview. Samantha's first question was, "What happened to Henry?"

At age twenty-seven, Henry was experiencing several epileptic seizures a day. He could have fallen down a stairwell and killed himself. The possibilities for catastrophic accidents triggered by a grand mal convulsion were endless.

In 1953, Dr. William Scoville, a neurosurgeon in Hartford, Connecticut, surmised that the hippocampal region of Henry's brain was the source of his seizures. With consent from Henry and his parents, Scoville decided to remove that trouble-spot. He drilled small holes in Henry's skull above his eyes and inserted thin metal tubes into the medial temporal lobe near Henry's midbrain—he could tell exactly where they were with X-rays (see Figure 1.1). He then carefully removed about half of the hippocampus via suction surgery. Also inadvertently destroyed was the amygdala, a structure near the hippocampus that is associated with emotion, especially fear. However, the best available evidence then and now indicates that Henry's cortex was virtually intact. Targeting a specific subcortical brain structure while leaving the cortex undamaged explains the scientific significance of Scoville's operation—the first of its kind. Equally important, the surgery effectively cured Henry's epilepsy and may have saved his life.

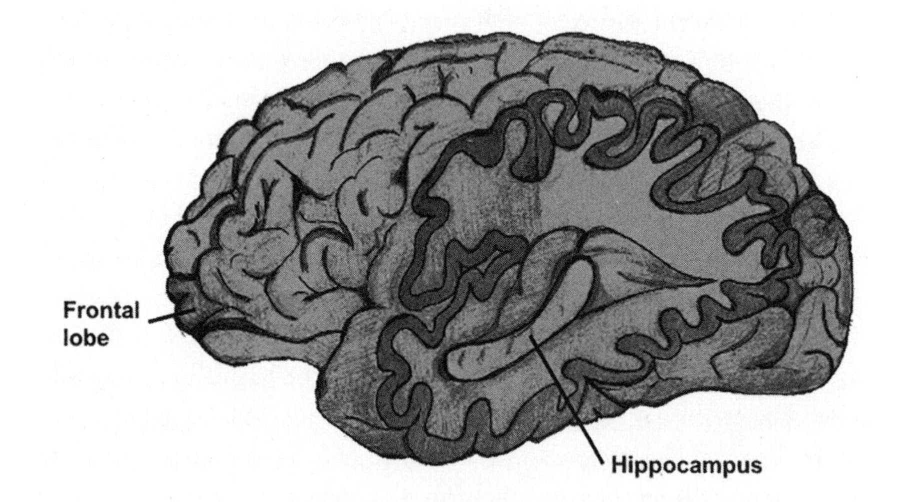

Figure 1.1. The human hippocampus in the medial temporal lobe next to the midbrain. (Artist's rendition of the original illustration by Henry Vandyke Carter appearing in Henry Gray's 1918 *Anatomy of the Human Body*.)

Soon after Henry's surgery it became obvious that something was terribly wrong. He could no longer remember things he had done hours, minutes, or even seconds earlier. Subsequent research on Henry's deficits revealed that the surgical team had inadvertently removed Henry's engine for forming new memories. Scoville quickly put out the word to other MDs: "Never repeat this procedure."

Henry's brain and behavior soon became a source of intense scientific scrutiny. Well looked after at a nursing home near Boston, Henry devoted the remaining years of his life to serving science. He participated in hundreds of psychological experiments at the Massachusetts Institute of Technology (MIT). He bequeathed his brain to MIT when he died in 2008 at age eighty-two.

Samantha questioned the value of research with Henry. Her concern foreshadowed some controversial issues raised in the 2016 book by Luke Dittrich, *Patient H.M.: A Story of Memory, Madness, and Family Secrets.*[3] One reader of Dittrich's book compared Henry's fateful operation to the procedures that Nazi brain surgeons performed on concentration-camp victims without their consent—just to see what would happen. I disagree with that comparison. The Nazis violated the Nuremberg Code that outlawed unwanted experimental surgery on humans. Scoville did not. He obtained the appropriate consent to perform Henry's operation. The Nazis were malicious. Scoville was not. He hoped to cure Henry's life-threatening epilepsy, and largely succeeded. In 1953, nobody knew that Scoville's surgery would shatter Henry's memory. No scientific evidence at the time indicated that the hippocampal region was critical to forming new memories.

Another reader of Dittrich's book compared postsurgery research with Henry to experiments performed by Nazi psychologists on Jews for the sole benefit of *Luftwaffe* pilots. This is an inappropriate analogy. Henry freely chose to participate in psychological experiments and he benefited immensely from that decision. He enjoyed, in his words, "helping science." Researchers in return ensured that he did not end up in a back ward of a psychiatric institution, forgotten and neglected.

Contributing to human knowledge also added meaning to Henry's life. He was able to interact with some of the sharpest minds on the planet. Young scientists with twice his education looked up to him. They promised Henry that what they learned about him would "help others."

Remembering tries to make good on that promise. Henry is on a continuum with normal people. Like Henry, we all sometimes misunderstand, misperceive, misremember, and think or act inappropriately. The book describes insights into Henry's catastrophic memory impairments, cognitive deficits and compensation strategies, together with what they indicate about how the normal human mind functions and occasionally breaks down in everyday errors. These insights can help the rest of us understand our cognitive strengths, put our periodic errors in perspective, and offset our occasional memory failures.

Samantha wanted specifics. "Based on your research with Henry," she asked, "can you please tell the readers of my newspaper how the normal brain forms memories?"

I responded by describing how *my* normal brain learned her name, Samantha Kimmey. I already knew the word *Samantha*, but *Kimmey* was new to me, so I had to link the syllables *kim* and *me* to form a *Kimmey* unit. I then joined *Samantha* and *Kimmey* to form a new cortical unit representing her full name.

My hippocampus played an essential role in creating this new *Samantha Kimmey* unit. The hippocampus is an activating mechanism that hyper-activates connected units—for time periods measured in seconds, a virtual eternity in the brain. Once hyper-activated, my *Samantha* and *Kimmey* units burned connections into a new cortical unit that served to represent that particular combination of words. I then attached everything I learned about Samantha to that new unit: where she is from, what she looks like, where she went to college, and her current job. This process is called *binding*. If I continue to use that *Samantha Kimmey* unit, its connections will gain strength and become long-lasting—a long-term memory. However, newly bound units must be used, or their weak new connections will disintegrate (see Figure 1.2).

Samantha was fascinated. "How did you become interested in memory, mind, and brain?" she asked. I described how a chance encounter at age twelve sparked my curiosity. An undergraduate just home from university had eagerly passed on to me some "revolutionary" facts about memory and mind that he read in *The Psychopathology of Everyday Life*. Freud! 1901![4] I smiled. Samantha laughed.

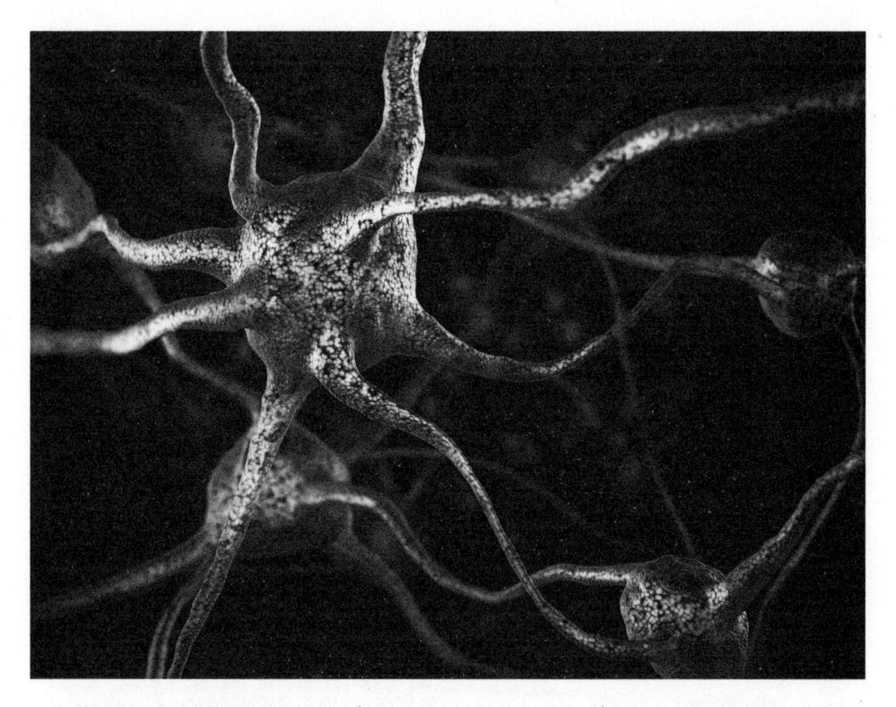

Figure 1.2. Connections between neurons in the cortex store memories. (Image by imageBROKER / Alamy Stock Photo.)

I also explained how my uncle Hank, a physiology professor, ignited my teenage interest in the brain. Inspired, I later enrolled in a graduate program at MIT, now known as the Department of Brain and Cognitive Sciences. In my second year there, my mentor, the department chair, Hans-Lukas Teuber, introduced me to Henry Molaison, soon to become the most famous amnesic patient in the world. This meeting was pivotal. I was witnessing the creation of a revolution in the understanding of memory, mind, and brain. My interest in those topics shifted into high gear for the remainder of my career at MIT and UCLA.

Samantha wanted more: "Tell me about the day you met Henry." That day is etched in my brain. In 1966, long before Henry became famous, my mentor delivered Henry to the office that I shared with another MIT graduate student. This was remarkable because I usually met Professor Teuber in his office (see Figure 1.3).

Figure 1.3. Chair of the MIT Psychology Department, Professor Hans-Lukas Teuber, circa 1969 in Los Angeles.

Squeezing into the doorway of my tiny room, Dr. Teuber introduced us as Don and Henry, as if we might become friends. Henry's thin, rather handsome face broke into a smile as we shook hands. He was then about forty years old, seventeen years my senior. I think I called him "sir." It never crossed my mind that this quiet man would become an international celebrity and a major focus of my research over the next half century.

Our three-way meeting lasted less than a minute. Professor Teuber excused himself to return to running the world-class psychology department he had built. He assured Henry that he would enjoy my experiment on sentence comprehension, "something you are good at."

Samantha asked how I became interested in language comprehension. Perhaps it was where I lived as a boy, a town in northern Quebec that was and perhaps still is fifty-fifty French-English. Canadians are still negotiating the fifty-fifty. Samantha laughed again.

I later decided to minor in language as a graduate student. The famous linguist, Noam Chomsky, was at MIT. He was revolutionizing the study of language and cognition. Psycholinguistics—the psychology of language—was a hot topic. I saw language as central to human memory, mind, and brain. It is built into our genes. It permeates human thought. It can influence how we see and remember things—a subfield unto itself.

"Did working with Henry change your views on language and memory?" Samantha wondered. I outlined some big changes. Because different types of memory follow different rules, I came to see memories as diverse but with an underlying unity. For example, my vision-based memory for a beautiful day knows nothing about the rules for conjoining adjectives and nouns to form a language-based memory such as *This was a beautiful day.* Henry showed me the unity underlying that diversity. His brain surgery impaired his language memories, his visual memories, and his event memories. The received wisdom at the time was that the hippocampal region only formed memories for events.[5] Not so. Henry taught me that hippocampal mechanisms help form many different types of memory, including the internal representations for perceiving and for communicating about the world.

Henry also taught me about amnesia. He was an amnesic, but we all begin life as amnesics. We forget events from our first three or more years of life, a phenomenon known as childhood amnesia. Some of us end life as Alzheimer amnesics. In between, we're just forgetters. Our brain is not as accurate, reliable, and well-behaved as a computer. Unlike computers, we misremember information, we forget information, and we relearn forgotten information.

My research with Henry even forced me to abandon an ancient philosophical intuition that enjoyed virtually universal acceptance at the time—that language and memory are separate and that memory means remembering events. However, Henry's operation didn't just compromise his ability to recall new events. It also impaired his memory for dozens of language categories.

"What are language categories? An example please," insisted Samantha. I returned to my memory for her name. Names and occupations are fundamentally different language categories in the brain. The information *This is Ms. Kimmey* is more difficult to learn than *This woman is a reporter*, especially for older adults, but also for young people. A reporter is an occupation. Kimmey is a name. People attach a whole network of associations to common nouns for occupations such as *reporter*, *baker*, or *carpenter*. They know what people in those occupations do, and they can easily link that information to a newly encountered face. But the support system is absent for proper names such as *Kimmey* or *Baker*. It's harder for people to link a new face with the isolated name *Baker* than with the otherwise identical word for the occupation *baker*. Names were likewise a special category for Henry.

Henry also taught me that memories are much more complex than anyone imagined at the time. Everyone thought event memories were simple. They're not. To form even the simplest event memory, the hippocampus must create and conjoin four units in the cortex for representing what event happened, where it happened, when it happened, and who experienced it. It is similar for language memories. When you initially learned the word *ungentlemanly*, your hippocampus conjoined the four units in your cortex that represent the components *un*, *gentle*, *man*, and *ly*.

My work with Henry showed that even reading aloud engages complex memory processes. English spelling only hints at how to pronounce most words. Readers must retrieve the correct pronunciation from memory. To read *ungentlemanly* aloud, they must segment its visual form into syllables and recall from memory that the syllable GEN receives primary stress. Syllabic stress is not in the letters, although it seems to be, because the retrieval process is so quick.

Henry's uncorrected reading mistakes showed how these normally quick retrieval processes can fall apart. Henry misread the familiar word *pedestrian* as "ped•AYE•ee•string." He could not retrieve the correct sounds for *pedestrian* from his frayed memory. He stressed the wrong syllable. He misread the second *e*. It's supposed to be short, not long. Henry's reading errors indicated that some of his word memories were rapidly degrading over time.

"Why do memories degrade over time?" asked Samantha. One factor is *disuse*. Words you rarely use suffer memory degradation, but not ones you use every day. If you have not used a word in months, then it's vulnerable to decay. Henry illustrated this vulnerability for *lentil*, a familiar word he rarely used. When asked for a definition, Henry said *lentil* was a "combination word" meaning "lent and till, area and time of." He no longer comprehended what *lentil* meant.

Aging is another factor. Memory degradation increases as normal people grow older. Difficulties in spelling, word-finding, and word-comprehension gradually worsen even though vocabulary usually continues to grow until the late seventies. However, Henry's difficulties worsened with aging at an abnormally rapid rate. Normal people can relearn and restore their memories for words degraded by aging and disuse, but Henry was not normal. He could not relearn forgotten words because his neural regions essential for new learning had been removed.

To illustrate the *combined* effects of aging and disuse, I told Samantha about my frustrating tip-of-the-tongue experience as she began our interview. Dave Mitchell, the editor of her famous newspaper back in 1984, was a friend of mine, but thirty years later, his name escaped me. However, the name came back to me during our interview, first the letter *M*, then *Mitchell*, and finally, *Dave*. Aging combined with a prolonged period of disuse had delayed retrieval of a name that once came readily to mind.

I then surprised Samantha with a question: "Can you guess the best way to *prevent* memory degradation?" It's *exercise*. Regular moderate exercise increases the blood flow to the brain and facilitates release of a protein known as *Brain Derived Neurotrophic Factor* (BDNF). By strengthening the synaptic links between neurons, BDNF may shore up the billions of memories stored in the brain of older adults. Exercise is especially important for older adults who want to keep their memory sharp. It's a hot research topic right now.

"Is memory decay inevitable?" Samantha wondered. Normal individuals can't control the fundamental causes of memory degradation—namely, aging and disuse. But unlike Henry, we can control *relearning*. After completely forgetting something, we can relearn it. Relearning revivifies a memory and resets the degradation clock.

Something else we can control is *engagement*. Keep active. Keep trying to remember. Take part in intellectual discussions. Stay socially connected. An active approach to life can help strengthen vulnerable memories. Your memories for everyday words such as *he* are not vulnerable. By one estimate, you will use *he* in its different forms—*his*, *himself*, *him*—at least twenty million times over your lifetime. You only have to worry about rarely used words.

You also control how much and how often you exercise. I get up from my computer, stretch, and take a short walk every hour and a half. I'm in aqua-aerobics groups. I'm in hiking groups. We talk as we walk along. Exercise and social engagement have toned up my body, mind, and brain.

You can even control what happens when memory fails. My wife, Dr. Deborah Burke, distinguished professor of psychology in the Linguistics and Cognitive Science Department at Pomona College, studies the tip-of-the-tongue phenomenon. If you are a participant in one of her experiments, she might ask you to retrieve a low-frequency word that fits a definition such as "a fastener that's made out of nylon." You think, "Oh, I know that word. It starts with *V* and has two syllables." Words that are not quite right might pop into mind, but you reject them. You return to trying to recall your largely forgotten word. My wife's experiments show that repeated attempts at recall may improve your ability to correctly recall a word up to a week later. Unsuccessful efforts to retrieve a lost word seem to be as effective as successful recall.

"Like when you were trying to remember Dave Mitchell's name?" Samantha suggested. "That's right. I recovered that name on my own," I answered. I did not consult my cell phone or ask her who edited her newspaper back in the 80s. According to my wife's data, my persistent attempts greatly strengthened my *Dave Mitchell* memory. Nobody really understands why. Many factors must be explored to determine why failed attempts to recall help later recall. Researchers all over the world are looking at that.

Samantha returned to Henry. "Why weren't his language difficulties immediately apparent to everyone?" she asked. One reason was that Henry's listeners were unaware of his errors during casual conversation—they automatically corrected them. Then too, Henry's language problems were complex. In everyday discourse, he retrieved common words and clichés without difficulty. He readily told and retold familiar stories from before his operation. He only became incoherent when forced to talk about something new. To get Henry to do that required ingenious experiments, plus a tape recorder to catch his errors.

In one of my experiments, participants—Henry and controls with no known brain damage—were instructed to use two or three prespecified words in a single grammatical sentence that accurately described a picture. To illustrate, I showed Samantha a sketch of a man on a street corner talking to his young sons and pointing at a *Do not walk* sign. I then asked Samantha to describe the picture using the words *first*, *cross*, and *before*. After a moment's thought, she said, "Boys, first you must look both ways before you cross the street." Her grammatical sentence contained all three target words.

Henry could not do that. This is his description of the same picture: "Before at first you cross across." His sentence included the three words but was completely incoherent. What he did was retrieve familiar words and phrases via free association: *at first*, *you cross*, and *across*, which is closely associated with *cross*. The target words triggered those existing associations in memory. He wasn't creating anything new to describe the picture. He just tossed his old memories into an ungrammatical salad.

"Did Henry know he wasn't making sense?" asked Samantha. In fact, Henry had no idea that he made errors. He lacked coherent internal representations against which to judge whether his utterances made sense or not. *Normal* people usually know when they make an error because it does not match their intent—the new and usually short-lived internal representation of what they intended to say. Unable to construct new internal representations in his brain, Henry could not sense the mismatch between his intent and an error such as *I want some her.*

Henry was also unable to detect errors that we planted in sentences and pictures and asked him to identify. In the many experiments I devoted to testing Henry's ability to discover errors in never previously encountered pictures and sentences, his error detection was always extremely impaired.

Samantha was curious. "How many experiments did you do with Henry?" Over my career, I ran about twenty-five experiments on Henry's language comprehension, visual cognition, sentence production, word knowledge, reading aloud, creativity, and sense of humor.

"What was your favorite" asked Samantha, "the most important, informative, or surprising experiment?" I replied that the picture description test she took was the most important and informative. The most surprising, however, was my first experiment with Henry as a graduate student, an experiment testing sentence comprehension. I discussed Henry's poor performance in that experiment at a lab meeting the next day.

In attendance were my mentor and a distinguished international panel of neuroscientists. They all considered Henry a pure memory case. For them, events were the only experiences that laid down memory traces. They did not see that creating a novel sentence involves new memory formation. They had not looked closely at Henry's speech errors. They assumed that his language abilities were normal.

Coming from a twenty-three-year-old graduate student, my lab talk was not well received, especially my conclusion: "Henry's sentence comprehension is abnormal. So is his language production. He's incoherent and has word-retrieval problems." Disapproval was swift and consequential. My advisor invited me to a mandatory meeting in his office.

I changed my thesis topic. Only as a UCLA professor many years later did I re-examine Henry's sentence comprehension results. By then I had abandoned the brain-first approach that I had lived and breathed as a graduate student at MIT. The most important step in that approach was to determine what regions of Henry's brain were damaged, *especially post mortem*. I left that to others. I preferred a feat-first approach, where the primary goal was to understand the everyday feats easily accomplished by the normal brain but not by Henry's. With this approach, my first questions were, *How do we ordinary folks understand novel sentences and visual scenes? How does our brain acquire new facts and remember recent experiences?* My next question was, *How do those everyday feats break down as a result of brain damage?* As I soon discovered, this feat-first approach was as unwelcome in the brain-first world as my earliest results with Henry—at least until recently.

Samantha wanted to hear more about my first fateful comprehension study with Henry. After nearly five decades, that study was still vivid in my memory. I remember thinking that my experiment would simply confirm my mentor's belief that Henry's language skills were normal. A wasted hour, then back to serious work.

But Henry surprised me. I gave him ambiguous sentences one at a time on cards and asked him to quickly find and describe their two meanings. An example is, *I just don't feel like pleasing salesmen*. I had expected him to say, *It means either I don't want to please salesmen*, or *I don't want agreeable salesmen around.* That was the typical response of Harvard undergraduates to the same sentence. Henry was different. He responded, "The person doesn't like salesmen that are pleasing to him. Uh, and that personally he doesn't like them and and [*sic*] personally he doesn't like them [*sic*] and then I think of a phrase that he would say himself, he doesn't, uh, pleasing, as conglamo [*sic*], of all of pleasing salesmen." Henry never did get the second meaning. And what he said was incoherent and incomprehensible.

"Does what you've discovered have any practical applications?" Samantha asked. "My friends warned me you might ask that question," I said. They wanted me to emphasize that pure research is important *without foreseeable applications*. Einstein is a good example. He invented relativity theory in 1905. Atomic energy was unimaginable at that time. His theory was *untested* for almost twenty years. Its applications came much later and were *surprising*. Benjamin Franklin made a similar point about a hundred years earlier. When asked, "What is the use of theory?" he replied, "What's the use of a newborn baby?"

Nonetheless, my findings *did* carry real world implications. I reminded Samantha how my research showed the importance for older adults to *use* their memories in order to prevent degradation. This *use principle* also applies to the hippocampus: use it or lose it. Keep active. Keep learning new information. Continue being creative. Engage in social activities. Focus on growing. Solve challenges, including the challenge of meeting new people and expressing new ideas.

Case H.M. illustrates what terrible things can happen when hippocampal damage prevents someone from creating and recreating memories. Henry's disaster, in a sense, had a silver lining. We learned from Henry how normal people can keep their memories strong.

Samantha had one last question. "If Henry had never had that operation, would the scientific community know as much about memory as it does today?"

Case H.M. was a major earthquake that forever reshaped the intellectual landscape of memory, mind, and brain.[6] Scientific understanding of memory took off with this one case. Even now, as I write this book, Henry's shattered memory and mind are shedding new light on what *normal* memory is, how it works, its role in creative expression, in humor and artistic endeavors, in perception and consciousness itself—everything that makes a human mind and brain worth having. *Remembering* is about this new knowledge and its implications for successful memory formation and retrieval at every stage of life.

Finally, some notes on how to read this book. *Remembering* is neither a biography nor an autobiography. It portrays in chronological order neither the events in Henry's life nor the year-by-year progression of my experiments with Henry. For readers interested in chronology, the Appendix at the end of the book gives a timeline of my research with Henry and the major events in his life.

Remembering will appeal to three overlapping types of readers. Some will want to learn about the new scientific studies and facts about Henry that have transformed our understanding of memory, mind, and brain. If you are one of these readers, I urge you to read the **ILLUSTRATION BOXES** interspersed throughout the book. These boxes enliven the facts and observations that interest you and the font of their titles (called **Broadway**) is intended to facilitate remembering (like illustrations in general) and to illustrate the concept of illustration. New and unusual fonts have been shown to improve recall of verbal materials[7] and the distinctive connotations of the Broadway font (show time in glitzy New York city between the two great wars) resonate with the memory functions of illustrations and with the core concept of *Remembering.*

The second type of *Remembering* reader will desire to understand their own memory and how it works. As one of these readers, the epigrams that begin each chapter can help you actively organize and integrate what already know with the information you acquire in a chapter. To illustrate, consider the Chapter 1 epigram, "Why do we forget the names of cherished friends we have known for many years?" Thinking about that question as you read Chapter 1 will help you organize, understand and remember the many facts therein that connect directly or indirectly with that epigram.

The third type of *Remembering* reader will want to maintain and enhance their memory, mind, and brain. If you are one of these readers, I urge you to answer the questions in the sections labeled *Test Your Memory* and *Questions for Reflection* that end each chapter in *Remembering*. They are not textbook or book-club questions. By answering the *Memory Questions* you will consolidate your understanding of the information in the chapter. By thinking about the *Reflection Questions*, you will further consolidate that information and perhaps also enhance your ability to remember the new information that you encounter every day. Please pay special attention to *Reflection* questions marked **. They are designed to help *you* specifically improve *your* memory.

Of course, the interests of many *Remembering* readers may intersect with all three reader categories. As one of these readers, I invite you to skip or skim nothing and to enjoy every facet of *Remembering*. You will discover revolutionary new information about Henry, *and* about your own memory and how it works, *and* about how to maintain and enhance your memory, mind, and brain.

QUESTIONS FOR REFLECTION: CHAPTER 1

1) Suppose your doctor advises that your hippocampus should be removed. After reading this chapter, what can you say to convince her or him that this is a bad idea?**
2) The author begins this chapter by describing a personal tip-of-the-tongue experience. Can you recall a similar experience that you have had? Did you eventually retrieve the correct word? How did you do it? How long did it take? **
3) Why do you think that you do not remember your first few years of life as an infant? How might childhood amnesia be related to how the hippocampus forms new memories?**
4) When asked "What's the use of theory," Benjamin Franklin replied, "What's the use of a newborn baby?" What did he mean?

TEST YOUR MEMORY FOR CHAPTER 1

1) What is the binding function of the hippocampus?
2) What radical new perspectives on memory and how it works did research with Henry inspire?
3) Why is a proper name like *Samantha Kimmey* so much harder to learn and recall than a common noun like *news reporter*?
4) What are some factors that cause memories to degrade over time?
5) How can you best counteract memory degradation?**

MEMORY TEST ANSWERS: CHAPTER 1

1) p. 14
2) p. 17
3) pp. 17–18
4) p. 19
5) p. 20

Chapter 2
WEIRD NEWS, INADEQUATELY RESEARCHED

HOW CAN YOU PRESERVE YOUR LANGUAGE MEMORIES AS YOU GROW OLDER?

In 2011, I chanced upon an article from the *New York Times* archives. It was about someone who could solve challenging crossword puzzles but had no memory. Who was this someone? Henry M.! Two years after his death, the *NYT* was praising Henry's "competent" crossword-solving abilities.[1]

This was weird. By then, I had published decades of accumulated data indicating that, after his surgery, Henry could neither comprehend uncommon words, nor read them aloud, nor spell them correctly—skills essential to solving even *moderately* difficult puzzles with clues such as, *A 10-letter word that means "an oven with a turning spit,"* with *S in position five and R in position eight.* To solve that puzzle, Henry would have to understand, retrieve, and correctly spell the infrequently used word ROTISSERIE.[2] This was not likely, according to my data. He would also have to reject words that resemble ROTISSERIE in meaning but lack *S* and *R* in the correct positions in the crossword matrix. And to establish *S* and *R* as the correct letters in those positions, Henry have to know the Down words SUBROSA and CHIMERA—again unlikely, given my results.[3]

Something was amiss. I just didn't know what. Either my data on Henry's word memories were somehow flawed, or the *NYT* was mistaken. I needed to know which. A major theory hinged on the answer. As another problem, audiences for talks I delivered on Henry's language use were asking a disturbing question: how could Henry solve difficult crossword puzzles if his word knowledge was impaired?

Resolving this crossword enigma required serious detective work. The outcome would interest readers worried about word-finding failures in everyday life, the most common memory complaint of normal older adults. The conclusion might also help readers understand and maintain memories for significant facts and events in their lives. Crossword puzzles may seem unimportant, but focusing on them follows a well-trodden path in science: to discover the world in a grain of salt, to expose the basis of life in the humble fruit fly, and in this case, to unravel the mysteries of memory in the lowly word.

RESEARCHING THE CROSSWORD ENIGMA

To evaluate the *NYT* claims, I first went over what I had learned about Henry's word memories, beginning on the day we met in 1966. As he described the meanings of the ambiguous sentences I gave him in our first experiment, Henry misused words in ways that continued to haunt me long after I earned my PhD from MIT. As you probably do not recall from Chapter 1, this is how Henry described one of the meanings of the ambiguous sentence *I just don't feel like pleasing salesmen*: "I think of a phrase that he would say himself, he doesn't, uh, pleasing, as conglamo [*sic*], of all of pleasing salesmen." Normal people describing that same sentence meaning typically said, *I don't want to please salesmen.*

As a young researcher, I did not know what to make of Henry's *conglamo.* Did he mean *concatenation*? Or the similar sounding *conglomeration*? Neither word fits the concept *to please salesmen* that Henry was trying to communicate. And if Henry inadvertently blended *concatenation* and *conglomeration* into the meaningless *conglamo*, why did he fail to repair this obvious error?

As Henry's errors piled up in that experiment, it seemed as if something was destroying his word memories. I just couldn't imagine what that something was. At age forty, Henry looked too young to be experiencing severe word-finding difficulties. I could not see the connection between Henry's brain damage and his deteriorating memory for words that he would have learned in grade school.

When I later set up my lab at UCLA, I began a line of research on word knowledge and the brain that stretched from 1972 to 2009. During this period, I re-examined my 1966 notes on Henry's word-finding problems and read a book-length transcript of conversational interviews with Henry, tape-recorded in 1970 when Henry was forty-four years old.[4] Questioned about his experiences during childhood and grade school, Henry's answers contained strange mistakes that he failed to correct. He told of making a model airplane out of *bamboo* or "very like wood," as if he could not remember the word *balsa.* He described his school friends as "more eased," an expression that reminded me of his *conglamo* concoction in my 1966 experiment. If Henry blended together the common phrases *more relaxed* and *more at ease* to create *more eased*, why did he not correct this understandable mix-up?

WHAT WAS IT?

Was Henry suffering from the type of dementia that older adults dread? One might suspect Alzheimer's disease if Henry made mistakes with *common* words such as *spoon* or *tree*, but he did not do that.[5] For someone age forty, Henry's word-retrieval problems seemed premature, but not pathological. He only failed to remember *rare* words such as *balsa*, a difficulty we all experience occasionally.

Could it be schizophrenia, another brain-related disease associated with memory problems and incoherent speech?[6] Henry showed none of the classic schizoid symptoms—social withdrawal, hallucinations, and irrational thoughts that persist.

I also dismissed *depression*—another, much more common brain-based disorder with well-established links to impaired memory and language competence. When I met Henry at MIT in 1966, he seemed undepressed—happy even. As far as I could see, he lacked the typical signs of serious depression—intense mental anguish, persistent bad moods, occasional outbursts, indifference to the world, and chronic inability to experience pleasure.[7]

If Henry's odd expressions were non-pathological, could they be a form of creativity? This too seemed unlikely. Henry's cliché-ridden utterances came across as blundering and baffling rather than inspired or creative.

So what was it? At age twenty-three, I had little knowledge of amnesia and no personal experience with Henry's kind of memory failure. What escaped me at the time was the possibility that Henry's disintegrating word knowledge could be a consequence of his unique brain operation.

That idea came to me out of the blue many years later. Henry's untimely retrieval failures were an inadvertent side effect of his fundamental problem: inability to learn, represent, and store new information in his cortex. People with normal memory often forget aspects of rare words. Not *every* aspect, of course. Some aspects of words get used so frequently as to be unforgettable, for example, recalling how to purse one's lips when producing the many words containing the sound */p/*. Only *underused* aspects in rare words are vulnerable to forgetting, say the *osa* in *mimosa*, a Spanish-origin word for a flower and a cocktail that you rarely discuss. That *osa* cluster is readily forgotten because, as a speaker of English, you probably use few other words containing that sequence of speech sounds. Underused sequences can vanish from memory almost as completely as if they had never been learned.

Somewhat more common words such as *balsa* can also suffer the same fate. But wouldn't normal people simply *relearn* partially forgotten aspects of a word when they subsequently encounter it in everyday life? Relearning would re-establish the original memory and offset the process of degradation that becomes especially noticeable in middle age. Henry's brain damage would have prevented this normal relearning and rejuvenation process, transforming the usually minor lapses we all experience into major impairments in his ability to read, comprehend, and recall familiar but underused words.

My out-of-the-blue idea that the hippocampal complex counteracts memory deterioration via periodic bouts of reconstruction was revolutionary. It swept away the standard view, that forgetting reflects a passive, continuous, and irreversible decay process. It also suggested a simple solution to a hundred-year-old mystery in research on amnesia. By definition, amnesics have difficulty acquiring new information *after* their brain damage. So why does the brain damage that causes amnesia usually also impair the ability to *retrieve* memories formed many years earlier, a phenomenon known as *retrograde amnesia*?[8] Until I examined Henry's word memories, no one clearly understood this paradoxical phenomenon or the parallel impairments in learning and recall that normal adults experience as they grow older.

So far, however, my data only hinted at some possible gaps in Henry's word knowledge. To determine whether Henry in fact suffered abnormal fading of established lexical memories, I needed to compare his word knowledge with that of people like him who were not amnesic.

ABNORMAL FADING OF ESTABLISHED MEMORIES?

The key person who helped me establish that Henry's word knowledge was fading abnormally as he aged was MIT Professor Suzanne Corkin. In her 2013 book, Professor Suzanne Corkin described herself as "Henry's keeper." She was also the gatekeeper for anyone interested in conducting research with Henry.[9] In 1997, Dr. Corkin graciously allowed me to test Henry's word memories at the MIT Clinical Research Center. With the help of my postdoctoral fellow at UCLA, Dr. Lori E. James, now a professor of psychology at the University of Colorado, Colorado Springs, I wanted to determine whether Henry could define uncommon words such as *squander* and *squeamish* and whether he knew they were real rather than invented words. I also wished to see whether Henry remembered how to pronounce rarely encountered words such as *abacus* when naming pictures and reading aloud. Finally, I hoped to examine Henry's memory for the unusual or irregular spelling patterns in words such as *fiery* and *endeavor*.

Dr. James and I first collected a set of words that Henry had known and used when interviewed in 1970. My lab then categorized the words as high- versus low-frequency based on how often people are likely to encounter them in print.[10]

We next inserted our words into five standard tests of word knowledge, and Dr. James flew to Boston to meet Henry and give him our tests at MIT. When Lori returned to UCLA, my lab analyzed Henry's data. Henry's knowledge of high-frequency words was normal on all five tests, consistent with impressions of the many people who have engaged Henry in casual conversation, including myself when we first met. However, I was now interested in Henry's knowledge for uncommon words such as *lentil*. For these words, Henry's performance was extremely abnormal on all five tests.

Testing Henry's word knowledge was the easy part. Much more difficult was the task of recruiting people who resembled Henry in age, education, socio-economic status, background, and IQ *but had normal memory*. To find these rare individuals suitable for comparison with Henry, we had to screen more than 750 older adults in the participant pools of the UCLA Cognition and Aging Laboratory and the Claremont Project on Memory and Aging. We also had to advertise for additional volunteers working in clerical or manual labor jobs in Southern California. Eventually, my team and I succeeded in identifying twenty-six individuals willing to serve as paid participants in a wide range of future studies with Henry.

We gave our word-knowledge tests to five memory-normal groups, each comprised of six to ten individuals. By comparing their results with Henry's, we discovered dramatic deficits in how Henry comprehended and used uncommon words. In our 1997 word comprehension test, Henry distinguished rare words such as *squeamish* from invented pseudo-words such as *friendlyhood* at chance levels of accuracy (50 percent), whereas the control participants averaged 82 percent correct. However, Henry scored over 86 percent correct when he took this same test in 1983. His normal comprehension level in this earlier assessment confirmed that Henry once understood the words he no longer recognized.[11]

Henry's ability to define uncommon words was likewise impaired, even though he used those same words appropriately when he was younger.[12] Unlike the control participants, Henry also produced erroneous definitions that we labeled *malaprops* and *misderivations*. Both types of errors highlighted the nature of Henry's word comprehension problems.

Malaprop definitions fit a similar-sounding word better than the target word, as in this excerpt, where Henry's definition of *squander* more closely resembles the similar-sounding *plunder* in meaning, the essence of a *malaprop* error:

Experimenter: What does *squander* mean?
Henry: Uh . . . to take things as one's own, other persons' things.

Misderivation definitions decompose words into fictional subcomponents, as in this excerpt. Note that Henry's definition dissects *lentil* into the fictitious subunits, *area* and *time* (*till*)—the essence of a *misderivation* error.

Experimenter: What does *lentil* mean?
Henry: Area and time of.

Participants without brain damage typically defined *lentil* as "a type of bean or pea." Henry, on the other hand, was trying to use the make-believe subunits *lent* and *till* to define *lentil*, in the same way that the actual subunits *un* plus *fortunate* define *unfortunate*. Even so, Henry's *lent* plus *till* don't really add up to "area and time of."

Henry's *malaprop* and *misderivation* errors in this study indicate abnormal deterioration of his memories for the meaning of low-frequency words, a conclusion reinforced in a two-part follow-up to the original experiment. In part one, Henry saw actual low-frequency words intermixed with pseudo-words such as *pediodical* and *reversement*. He was directed to respond *yes* for words and *no* for non-words. In part two, we asked Henry to define all of the words and non-words he earlier claimed were words. The results showed that Henry produced similar *misderiva-*

tion and *malaprop* definitions for words and pseudo-words, confirming that his near chance level of performance in part one reflected guessing. Like his malaprop definition for the real word *squander*, Henry used sound-similarity with *periodical* to define the pseudo-word *pediodical* as "about the same thing as *periodical.*"[13]

Henry also produced hundreds of spectacular errors in subsequent studies that asked participants to read rare words aloud. As three typical instances, Henry misread *triage* as "triangle," *thimble* as "tim•BO•lee," and *pedestrian* as "ped•AYE•ee•string." The control participants we later tested produced many fewer reading errors, and none as bizarre as Henry's.

We reasoned that the peculiarities of English spelling combined with amnesia contributed to Henry's remarkable reading mistakes. To correctly read aloud a multisyllabic word such as *pedestrian*, one must retrieve from memory the pronunciation features not specified in written English: the number of syllables, the syllable boundaries that, for example, call for saying "pe•des•tri•an" rather than "ped•est•ri•an" or "pe•de•stri•an," and the pattern of syllabic stress for saying "pe•DES•tri•an" rather than "PE•des•tri•an" or "pe•des•TRI•an." One must also recall the idiosyncratic ways we pronounce certain letters such as the *e*'s in *pedestrian*—which are short, as in *pet*, rather than long, as in *beat*. Most readers are unaware that these features are unwritten—they simply remember how to pronounce the words. Henry's degraded word memories, however, were not helping him fill in the unwritten information.[14]

WHAT'S IN A NAME?

Henry's problems were similarly apparent when asked to name line drawings of familiar objects on the Boston Naming Test. If participants could not recall a name, say, *trellis*, the experimenter provided a semantic cue, "it can support a plant," then a pronunciation cue, "it begins with *tr*," and finally, the word itself in a verification question, "Do you know the word *trellis*?" On this test, Henry correctly named many fewer pictures than control participants his same age, benefited less than others did from pronunciation cues, and produced significantly more erroneous answers that contained fewer than 30 percent of the correct speech sounds, as when he called a SNAIL a "sidion."[15]

It wasn't that Henry had never encountered the depicted objects or failed to learn their names during childhood. He usually recalled *some* of the target sounds, together with additional information about the object, as when he identified a PALLETTE as "like an easel." These remnants of Henry's severely degraded word memories further confirmed that he once knew the names he could not recall. So did his responses to the verification questions, as this exchange illustrates: [16]

> **Experimenter**: Do you know what a stethoscope is?
> **Henry**: Well a doctor uses it to uh find out different areas of you. They . . . uhh, yeah. Or how your heart is working and th-the noises go coming from that [inaudible] goes to other parts.
> **Experimenter**: Do you know what asparagus is?
> **Henry**: Yes . . . Cause uhh we had it in our soup my mother had learned how to make, the soup for the, my father, down south.

In our spelling task, participants heard a familiar but irregularly spelled word such as *bicycle* while seeing it spelled with a missing letter, here, BIC_CLE. Below this were two letters (in this example, *I* and *Y*), one of which could correctly fill in the blank. Our comparison group chose the correct letter for significantly more words than Henry, who again performed close to chance (50 percent). Just as disuse had eroded Henry's once intact memories for the meaning and pronunciation of rarely encountered words, so too with his memories for uncommon or irregular spelling patterns.[17] Like most other amnesics with deficits in forming *new* memories for information encountered *after* the onset of amnesia, Henry was also experiencing something familiar to normal older adults: *retrograde* amnesia or problems in remembering facts learned long ago. The difference was that Henry's retrograde forgetting was much more severe than normal.

WHAT'S OLD CAN BECOME NEW AGAIN

Henry's shredded memories for rarely used words fit my original idea.[18] Unlike normal individuals, Henry could not *relearn* completely forgotten information so as to correct the deterioration of memory that accompanies infrequent use. His hippocampal lesion had destroyed his ability to represent new information in his cortex—and to reconstruct or *re-represent* information that disuse had degraded. As a result, ordinary word and spelling retrieval problems escalated for Henry into a serious inability to read, comprehend, and produce familiar words.

If correct, this simple idea readily explains the hundred-year-old mystery of retrograde amnesia: underused information that Henry and other amnesics with hippocampal damage learned *before* their brain trauma is especially difficult to remember, not because amnesics forget at a faster than normal rate, but because, unlike normal individuals the same age, they cannot *relearn* information they have forgotten *after* their injury.

DATA SHREDDING AND THE CROSSWORD ENIGMA

In 2001, 2002, and 2009, neuroscience journals such as *Hippocampus* published my strikingly consistent data on Henry's deficits in comprehending, reading, and spelling words, together with my relearning hypothesis. Convinced that Henry's word memories were genuinely impaired and that his deficits made sense in light of his inability to recreate faded word memories, I revisited the only remaining hypothesis for explaining the crossword enigma: the *New York Times* made an error. Contrary to the *NYT* article, Henry was not a competent solver of crossword puzzles after his operation.

The *NYT* had reported that Henry could solve crossword matrices unsolvable by others, that mastering challenging word puzzles gave him "great satisfaction," that he developed his "crossword puzzle habit" at a young age and completed two or more puzzles a day beginning "in his teens," that he filled in "books upon books" of puzzles over his lifetime, and that he always kept a pen and puzzle book "with him, morning, afternoon, and night," even near the end of his life.[19]

To verify this information, I turned to the *NYT*'s primary source: a 2008 research report that praised Henry's "erudite" knowledge and use of words when completing crossword grids.[20] A team of Duke University researchers coauthored the article, which appeared in the British journal *Memory*. I expected at least 90 percent overlap between the answer keys and Henry's solutions—a reasonable definition of competent that would have supported the *NYT* claim.

I was disappointed. The researchers did not report percent correct. They didn't even indicate how many clues Henry attempted to solve in the 277 puzzles in the six books they analyzed. They only reported that Henry made 2,834 errors. The *NYT* had called 2,834 mistakes "competent" without asking for the obvious measure of competence—percent correct puzzle solutions!

Of course, I could easily compute this percentage on my own if I could obtain a copy of Henry's puzzles. I could not, however. The Duke University team refused a formal request for them, and so did Dr. Suzanne Corkin, Henry's keeper at MIT, as you may recall. These refusals to share copies of Henry's puzzle books violated the publication guidelines of the American Psychological Association (APA): that researchers must safely store their data for five years following publication and share them with competent researchers who request the data and offer to share the costs of compliance. At that time, however, the APA lacked procedures for disciplining violators of its data-sharing rules.[21]

So I could only evaluate Henry's puzzle proficiency by delving deeper into the article published by the Duke University team. There I discovered something curious. Henry misspelled many words in his puzzles, but the Duke team coded them not as misspellings, but as "alternative" responses that they considered superior to the answers in the keys! And they failed to provide a full list of those "alternative spellings." Instead they gave only a single example with a five-letter clue in the context _ _ _ _ _ *Rica*. The correct answer in the key was *Costa Rica*.

Henry had written *Porta (Rica)* instead of *Costa (Rica)*. As the Duke team noted, Henry had misspelled *Puerto* as *Porta*, which they accepted as correct because they believed Henry was creatively misspelling "common words to fit the constraints of the crossword puzzles." However, nobody on the Duke team analyzed the use-frequency of Henry's misspelled words. *Puerto* is *not* a commonly used word, and, as I soon discovered, Henry paid no attention to vertical constraints in crossword puzzles.

In our experiments, Henry misunderstood, misread, and misspelled uncommon words but not common ones. To contradict our results, the Duke researchers would have needed to analyze the use-frequency of the correct responses in Henry's puzzles. They did not do so, probably because it would have been pointless: The puzzle books they examined were for *children*. No knowledge of rare words was needed![22] This is the clue for a typical Across word in one of those puzzle books: *A seven-letter word that means "an illustration or typical case,"* with *A in position three, P in position five, and E in position seven.* To establish *A*, *P*, and *E* as the correct letters for those positions, Henry had to know the Down words SPREAD, COUPLE, and INVENT. The correct answer in the key is EXAMPLE.

A second team led by Duke University researchers alleged that, according to those "close to him," Henry solved "challenging puzzles featured in books published by the *New York Times.*"[23] Whoever those close people were, they must have been embellishing. As illustrated at the start of this chapter, Henry almost certainly could not solve *moderately* difficult puzzles. That Henry could solve the famously difficult puzzles in *NYT* books is extremely unlikely. However, we will not know for sure until those close to Henry make all of his filled-in puzzle books available to qualified independent researchers.

What I discovered next was truly shocking: Henry didn't really know how to do crossword puzzles. He violated the basic rules of the game. Rule one is to satisfy the constraints that Down clues impose on Across answers and vice versa. However, the Duke University Team reported that, *without exception*, Henry filled in Across answers, beginning with 1 Across, and only later, if at all, did he examine Down clues in the novice-level puzzles that he filled in at home.[24] Henry's puzzle "solutions" violated the basic concept of *cross*-word puzzles.

Henry also broke an even more important, *unwritten* crossword puzzle rule: solvers are not free to modify the clues to a puzzle. In the example provided by the Duke University Team, Henry responded *Porto Rico* to the clue _ _ _ _ _ *Rica.* This was doubly wrong. Not only did Henry offer a six letter idea (*Puerto*) for a five letter clue, *he changed the clue from Rica to Rico*. By altering the published clues to this and perhaps other puzzles, Henry was playing some other game. And so was the Duke University team that accepted Henry's *Puerto Rico* idea as correct instead of the answer in the key: *Costa Rica*.

These observations cracked the original crossword enigma to my satisfaction.[25] The usually accurate *NYT* was wrong about Henry's puzzle proficiency—a case of news inadequately researched rather than fake news. Our experimental data were solid. No evidence indicated that Henry could correctly spell rarely used words or that his word knowledge was erudite or that his crossword solutions were average for his age, let alone competent for someone his age.

This news should reassure readers concerned about word-finding failures in everyday life. Unlike Henry, normal individuals like you and I can easily *relearn* information we have forgotten. By using the intact hippocampal region of our brain, we can offset forgetting and keep our memories sharp. Memory-making can defeat memory-breaking.

A DETECTIVE STORY CONCLUDED: WHAT GAME WAS IT?

Reassured that my data on Henry's word memories were sound, I again pondered the *New York Times* article. Did Henry really solve crossword puzzles that others could not? The Duke University Team did not ask others to solve the puzzles that Henry worked on at home. There was no control group—a fundamental flaw. And when the Duke University Team compared the performance of Henry versus controls on crossword puzzles they constructed in the laboratory, the puzzles were so easy that performance was near ceiling for Henry and at ceiling for the controls. This means that the controls could not have outperformed Henry in principle on those puzzles—another fundamental flaw in the research.

Then there was the *NYT* claim that Henry kept a crossword book close by "morning, afternoon, and night" near the end of his life. How could he remember to do that? During this period, *dementia* as well as hippocampal damage was obliterating Henry's memories.[26] Perhaps it was someone else who remembered to lug the book and pen around with Henry—someone with normal memory. Were Henry's nurses instructed to keep his crossword puzzles within reach, much like the beloved teddy bear of a patient suffering from dementia?

Finally, the biggest anomaly of all. Is it true that Henry was "proficient" at solving puzzles "as a teenager," long before his neurosurgery made him amnesic? Amnesics quickly forget recent events, but they do not forget highly practiced skills that they continue to use after the onset of amnesia.[27] If Henry was an expert crossword puzzle solver during his teens, those solving skills would have remained intact after he became amnesic. At the very least, he would have retained the basic rules of the game. However, Henry was playing a different game after his surgery—one that did not require cross-checking Across answers with Down answers.

What game was it? What "great satisfaction" drove Henry to generate thousands of incorrect crossword solutions? Was he feeding his desire to help others? Henry tirelessly reiterated this wish to anyone willing to listen. Henry's determination to help also inspired him to participate in hundreds and perhaps thousands of experiments—making him the most studied patient in the history of the brain and behavioral sciences.[28] By pretending to master challenging word games, was Henry trying to help his keeper support her oft-repeated claim—that he had memory problems but no language problems? When he died in 2008, Henry may have carried this mystery to his grave: in December 2015, Luke Dittrich recorded Professor Corkin saying that she had shredded most of Henry's records, data, and transcripts that she had accumulated over many decades at MIT.[29]

Because Dr. Corkin apparently published some of those data as late as 2013, her alleged shredding may have violated the APA guideline that researchers must retain and make available copies of data for at least five years following publication. However, MIT has yet to confirm what data are missing (if any). Dr. James DiCarlo, chair of the Brain and Cognitive Sciences Department at MIT, inherited Henry's records after Dr. Corkin's death in May 2016, and in August 2016 he announced in public that none of Henry's files had been shredded. I was happy to learn this, and on the assumption that MIT could now make specific files available to competent researchers, I made a formal request in early October 2016 for a copy of Henry's crossword puzzles. Dr. DiCarlo wrote that he or his assistant would get back to me soon regarding this request, but now, four years later, I had not received those crossword puzzles, despite repeated reminders sent to Dr. DiCarlo via email and snail mail. It is difficult not to conclude that—despite Dr. DiCarlo's published statement to the contrary—Dr. Corkin did in fact shred some of Henry's important data files, including his crossword puzzle books.

QUESTIONS FOR REFLECTION: CHAPTER 2

1) What enjoyable activities besides solving crossword puzzles would you recommend for maintaining language memories as people grow older?**
2) Given that Henry's memory problems only applied to rare words, how do you think Henry would perform on nonverbal memory tests of his ability to identify rare sounds and uncommon objects?**
3) How would you feel if you kept forgetting words you once knew and used? Why didn't Henry panic when he kept forgetting words he once knew?**
4) Henry solved crossword puzzles incorrectly by focusing only on the Across words and by substituting made-up words for actual words. Do you think Henry was unaware that his solutions were incorrect, just as he was unaware of his other types of errors?**

5) As mentioned in the end of the chapter, practicing crossword puzzles will help maintain your memory for the particular words tested but won't necessarily help with other memories. Why do you think this is the case?**
6) This chapter discusses how the *osa* in *mimosa* is a rare letter/sound combination in English, and so it's much easier to degrade in memory if not used frequently. However, *osa* is a common syllable in other languages, such as Spanish. Given this example, what would be a benefit of bilingualism for linguistic memory and how would this work? Would there be additional benefits to knowing three or more languages or would that have some negative consequences?
7) Why do you think the published results of the Duke University team were so contrary to those of Dr. MacKay's UCLA lab?**
8) What examples from your own life could you use to help a friend understand what *retrograde amnesia* is?**
9) How can a normal individual such as yourself slow down how fast your language memories degrade?**

TEST YOUR MEMORY FOR CHAPTER 2

1) Why is your ability to *relearn* such an important defense against amnesia?**
2) Why couldn't Henry do what normal people seem to do: *relearn* completely forgotten information about words so as to correct the memory deterioration associated with disuse?
3) When asked what *lentil* meant in 1997, Henry said, "Combination word: lent and till, area and time of." What does this definition illustrate about Henry's memory for low-frequency words?
4) How did the peculiarities of English spelling help highlight the degradation of Henry's word memories?
5) Why do amnesics, who *by definition* have difficulty learning *new* information encountered *after* their brain damage, also have difficulty remembering facts learned long *before* the onset of amnesia?

6) Why were schizophrenia and depression considered possible reasons for Henry's language and memory problems?

MEMORY TEST ANSWERS: CHAPTER 2

1) p. 32
2) p. 32
3) pp. 34–35
4) p. 37
5) pp. 37–38
6) p. 31

Chapter 3

HOW CAN YOU HELP VULNERABLE MEMORIES SURVIVE?

ARE YOU MAINTAINING YOUR WORD MEMORIES?

With the crossword enigma off my desk, I turned my attention to the ordinary folks we compared with Henry in our word-knowledge experiments. What do Henry's collapsing word-memories say about the everyday word-finding problems of normal people?[1] An obvious way to prevent embarrassing retrieval failures is to shun the use of uncommon words. But favoring everyday words will not work in the long run. Every unreliable memory you dodge will become more difficult to retrieve the next time you need it. Eventually, the many valuable but underused words, facts, and concepts accumulated over your lifetime will require cumbersome stand-ins. By avoiding the word *zebra*, you may find yourself destroying a conversation by asking, "What are those horse-like animals with the black-and-white striped bodies?"

Active use and exposure is a better approach. Research shows that repeatedly trying to retrieve a forgotten word will increase the likelihood of success up to a week later, and once retrieved, you can reiterate the word to reduce the probability of subsequent retrieval failure.[2] Solving crossword puzzles can also help sustain your word knowledge—if you follow the rules of the game.[3] Actively participating in games such as Scrabble rather than passively watching television can likewise help preserve your memory for uncommon spelling patterns. To engage and restore even more memories, join a book club. I meet regularly with a friendly group of book lovers around a table in our local library. We discuss our book of the month for about an hour, a fun way to restore an enormous range of memories. Then comes the hard part: we each lobby for what book we will read next time. To win the votes of my fellow bibliophiles for my type of book, I must recall their names and what they like to read. No easy feat but a good workout for the brain.

If you like to write, joining an author's club can help repair many additional memories. After I decided to create this book, I joined the Claremont Authors' Club—a carefully chosen group of twelve word-slingers. Every week, each of us has the option of reading aloud up to two thousand words that we've written. The others generate constructive verbal comments and write suggestions for improvement on a copy of the work. Some readers get praise for their vivid metaphors. Others hear reminders of forgotten rules regarding coherence, precision, or punctuation. I often receive a jocular warning: "Lose the academic style. Most readers of your future book will be *people*, not scientists."

The ways to shore up underused memories are endless. Join a dance club if you enjoy movement and want to relearn forgotten routines. Connect with a Karaoke club if you love singing and want to refresh your memory for popular song lyrics. Volunteer to tutor students in high school English classes. You'll mend your memories for great literature. Offer to help immigrants learn English. You'll enjoy the psychological rewards of mentoring others. Or help your internal representations survive by forming a dining club. Once a week I lunch with an all-male group of intellectuals jokingly called ROMEOs: Retired Older Men Eating Out. Our open-ended discussions tune up our memories for jokes, philosophical issues, politics, ongoing international events, medical matters, and our long and colorful personal histories.

On a smaller scale, bring a friend when taking in a movie or art exhibit. Discussing your experiences afterward can restore a surprising number of slippery memories. I recently explored a vast multistory artwork with a friend.[4] A sign at the entrance to the exhibit read *Fascia* (*fash-ee-ah*), a word I once knew but no longer remembered. Inside was a network of paint-streaked sheets and ligatures that the artist invited us to move through, touch, and walk on. We saw a silent film in a tapered, dead-end passageway containing a soft cushion-bed that we lay on. We overheard an emotional psychotherapy session emanating from hidden speakers high overhead. I had no idea what it all meant.

My friend was also perplexed. Why the carefully stitched sleeves, canvas strips, and elongated pant-legs that connected with each other up to the roof and across the floor? What to make of the animated treasure chest containing a dancing pearl that was projected onto a monitor at the end of the narrow curving alcove? Why the teary dialog heard from high overhead as if in a dream?

The key to solving these mysteries lay hidden in my friend's cortex. She remembered what *fascia* are—sheet-like connective tissues that bind isolated muscles and sections of the human brain and body into groups, organs, and systems. Using the intact hippocampal region of my brain (see Figure 3.1), I was now able to form a coherent internal representation that made sense of everything we saw.

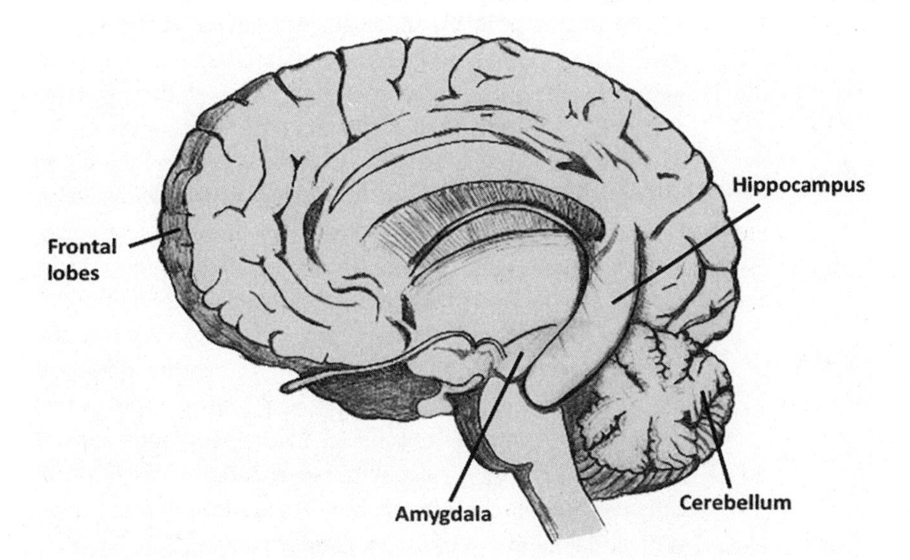

Figure 3.1. The hippocampal region of the brain. (Image by Patrick Phalen.)

We had taken a metaphoric stroll through the inside of the artist's body, ending up inside her womb! The swaying pearl was one of her human eggs! We had also toured her mind and the fragile social tissues that connect people together in friendships, families, and societies. We even eavesdropped on one of her psychotherapy sessions. Discussing that exhibit with my friend I relearned the word *fascia* and permanently cemented it in my brain. I wanted to congratulate, thank, and hug the artist, but by then, she had gone.

REFLECTION BOX 3.1: YOUR WORD-MEMORIES, AGING, AND THE AMERICAN MEDICAL SYSTEM

How are your word-memories similar to the current American medical system? For both, there are two ways to offset the effects of aging. One way involves slow incremental maintenance over decades. The other way involves reconstruction or building anew.

Heart health in the medical system illustrates this contrast. Primary care physicians and maintenance cardiologists help older patients nourish their heart through diet, exercise and pills. This slow, long-term process often continues for decades, but is highly effective in reducing heart failure and is inexpensive for society and for medical organizations alike. Maintenance cardiologists often get by without even a nurse, averting age-related disasters such as strokes and heart attacks at a fraction of the cost of emergency intervention.[5]

Invasive cardiologists are the MDs who perform the urgent surgery needed to replace or repair the damaged hearts of older adults, a risky process that is often unsuccessful and invariably expensive. Cardiac surgeons earn at least double the salary of maintenance cardiologists and typically require an operating theater, a squadron of highly paid staff, and equipment costing millions of dollars.[6]

The situation with your word-memories is similar. Routine maintenance is highly effective in combating effects of aging, but requires no specialized brain equipment, and costs almost no time or effort. Even conscious awareness is unnecessary. All you have to do is use the words you already know. The word frequency effect illustrates the benefits of use in its simplest form. Words that you use frequently rather than rarely over the course of your life are less susceptible to the ravages of normal aging, dementia, and brain damage (including Henry's hippocampal lesion).

By contrast, relearning words you have forgotten requires time, effort and the evolution of specialized brain mechanisms such as the hippocampus for creating new internal representations in your cortex (see Chapter 7) and for becoming aware of novelty and your own self-produced errors (see Chapters 17 and 20).

A caveat is now in order. Maintaining memories through use will not tune up your entire brain. The benefits of using a memory tend to be specific to that particular memory. Solving crossword puzzles may help restore your memories for how to spell and pronounce the rare words in the puzzles, and perhaps also some words with similar spelling or pronunciation. However, crosswords almost certainly will not help you remember where you parked your car or misplaced your house keys. Most studies in the vast literature on cognitive skill show virtually no transfer of learning to genuinely different information and situations.[7] The next chapter will discuss a much more powerful way to stave off memory loss in general and sharpen your brain across the board.

QUESTIONS FOR REFLECTION: CHAPTER 3

1) Has this chapter given you any new ideas for how you can maintain your word memories as you age? What do you already do to maintain your memories?**
2) How would an active social life help you restore word memories degraded by disuse or aging?**
3) How might learning another language help you retain words in your native language?**

TEST YOUR MEMORY FOR CHAPTER 3

1) Instead of avoiding the use of uncommon words, what are some better strategies you could adopt to preserve your memory for both common and uncommon words?**
2) Why is playing word games such as crossword puzzles a good strategy for preserving your word memories?**
3) Working to maintain your word memories does not guarantee memory improvement across the board. Why not?

MEMORY TEST ANSWERS: CHAPTER 3

1) pp. 47–48
2) p. 51
3) p. 51

Chapter 4

CREATIVE AGING: A SILVER LINING

They say that memory is the second thing to go with aging, but I can't remember the first.

—Anonymous

Back in graduate school, I learned that recently acquired memories for events gradually erode as time passes, with either decay or interference from similar, subsequently formed memories as the mechanism. Hermann Ebbinghaus, the German psychologist who pioneered the experimental study of memory, had established this in 1885. In the 1990s, I discovered that well-established memories for the spelling and pronunciation of words slowly become harder to retrieve as normal adults grow older. My research on the tip-of-the tongue and tip-of-the pen phenomena had shown that.

However, in 1997, something seemed to be wreaking havoc on Henry's word memories at an extremely rapid rate as he aged. Was Henry's hippocampal damage to blame? If so, this raised an interesting question about normal memory: can the intact hippocampal mechanisms of normal older adults create new memory components to replace the long-established ones damaged by aging? This question challenged the notion—prevalent since the 1960s—that normal aging continuously and irreversibly bleaches memories until they become permanently irretrievable.

As an *older adult* myself by current research standards (age sixty-five years or older), I know from personal experience that I often relearn information that aging—as well as disuse—has degraded. In 2014, I was absolutely certain that I met Henry in 1967. I nevertheless checked the date when proofreading my *Scientific American Mind* article. To my surprise, I was wrong! An unpublished write-up of my experiment with Henry on the day we met indicated it was 1966, a date I will not soon forget. Using my intact hippocampal mechanisms, I reconstructed that memory, counteracting the ravages of almost fifty years of aging and disuse. This is a common experience. When older adults discover they no longer remember a familiar name or event, they often look it up or ask someone about it. They then make a point of trying to remember it, often by repeating the information to themselves.

If the rebuilding and restoring of faded memories is a normal function of the hippocampal region, I needed to know whether Henry's memories followed an abnormal trajectory as he aged. And what better way to find out than to examine Henry's word knowledge as he grew older? The same basic principles of memory storage apply to all memories, but determining how aging affects memories is easier for word knowledge than for personally experienced events—which is what comes to mind when most people think about *memory*. For starters, children learn words in their native language at a fairly consistent age and use them with predictable frequencies in daily life. By contrast, personal experience of events can occur with unknown frequencies and at unknown and variable ages. Moreover, how often people use their memories for particular events in everyday life is virtually impossible to determine.

By examining the humble word, I hoped to unlock the mysteries of memory and aging, including the counterintuitive but well-established fact that amnesics forget events experienced in the days, months, and years immediately preceding the onset of amnesia at a faster than normal rate, but forget more distant events at a basically normal rate. I saw a way of solving this longstanding enigma if I could show that Henry's memories declined with aging and disuse at a faster than normal rate.

REFLECTION BOX 4.1: EMBRACE THE SLOW

Like the rest of the body, the brain and mind gradually slow down with aging. Let it be. What counts is your attitude toward speed.

Older adults in every society, regardless of gender, background, and religious conviction, share one thing in common: we slow down. All behavior is somewhat slower in older than younger people—a difference measured in thousandths of a second. As a recovering speed-alcoholic, I discovered that milliseconds count for little in the life of the mind. Slower doesn't always mean duller or less interesting. I've rethought my relation to time. I've slowed down the pace and I'm happier. I no longer try to compete with my ultra-fast computer and iPhone. I've rethought my relation to time. I've slowed down the pace, and I'm happier. You too can lose the speed-up mentality. Life is not a race. Savor the moment. Join the Slow Revolution. Time is just a benevolent aspect of the universe. You cannot save it. You cannot lose it. Time does not drain away. It is infinite.[1]

A TYPE OF FASTENER MADE OF NYLON

After setting up my lab at UCLA, I conducted a wide range of experiments on how aging impacts the memories that support the comprehension, spelling, and pronunciation of familiar words. My results over the years delineated specific age-related changes in word memories. In 1991, for example, my colleagues and I reported systematic declines with aging in the ability to retrieve familiar but rarely used words. In one experiment, adults age sixty-five and older couldn't recall as readily as eighteen-year-olds the word corresponding to a definition such as "a type of interlocking fastener made of nylon." For the older adults, *Velcro* more often remained "on the tip of the tongue," and in other experiments, reports of this tip-of-the tongue experience gradually increased in frequency with aging. This slow increase in the frequency and severity of retrieval failures for rarely used words suggested that aging was severing the existing links in memory between meaning and phonology—detectably so after age thirty-six!

All of the participants felt they knew the word on the tip of their tongue, and they often realized how many syllables it contained, and which one received greatest emphasis (VEL). Sometimes they even knew its first sound (V) without being able to retrieve the rest of the word. These *word fragments* more often came to mind for young than older adults.

The participants were right, however. They did know the word on the tip-of-their tongue despite being unable to bring its full pronunciation to mind. After the experiment, the young and older groups equally often chose *Velcro* as the sought-for word in a lineup of words with similar meaning and sounds, consistent with the fact that information is generally easier to recognize than to recall.[2]

Aging also impairs the ability to recall familiar but rarely used phrases, especially phrases consisting of a first and last name. You may recall from Chapter 1 my difficulty in recalling the name phrase *Dave Mitchell*. Despite knowing Dave well, I drew a blank on his name. I had not used that combination of words for many years. My problem was threefold: advanced age, disuse, and special difficulties with proper names—which are harder to recall than other types of words, especially for older adults.

However, I eventually did retrieve my *Dave Mitchell* memory. I didn't ask my interviewer or my computer, "Who edited the *Point Reyes Light* in 1980?" Instead, I tried repeatedly to recall the name on my own, and *Dave Mitchell* slowly came to mind, first *M*, then *Mitchell*, and finally *Dave*. Even when unsuccessful, repeated attempts at recalling a memory without help will improve your chances of success in the future.[3] And the name *Dave Mitchell* has stuck with me for many years now.

A HUNCH ABOUT AGING AND AMNESIA

My research team discovered the spelling analog of the tip-of-the-tongue phenomenon in 1997. We called it the tip-of-the-pen phenomenon. In a simple spelling test, older adults misspelled familiar but irregularly spelled words such as BICYCLE, RHYTHM, PHYSICIST, RHAPSODY, and YACHT reliably more often than young adults. These spelling difficulties increased systematically from age sixty-five to seventy-five and older. At age seventy-five, people who knew how to spell BICYCLE in grade school often could not remember whether BYSICLE, BISYKLE, or BICYCLE was correct.[4] Sometimes adults over age eighty-five could not correctly read the word *bicycle*. At this point, *bicycle* no longer occupied "the tip of the pen"—aging had severed the links in memory between the visual form of the word and its pronunciation.

These findings were decisive in developing my theory of aging, memory and forgetting. I now saw that degradation of neural connections in the neocortex accelerates with aging and underlies the tip-of-the-pen and tip-of-the-tongue phenomena. Links between speech sounds and spelling, and between the meaning and pronunciation of words that older adults rarely hear, read, say, or write, progressively weaken until retrieval becomes difficult, and eventually impossible. Frequent use or recent exposure, however, strengthens damaged connections and prevents or at least slows down the degradation process. Older people show no deficits in recalling information they often use in daily life.

Within this framework, I reasoned that normal adults may create new cortical connections to represent completely forgotten information that disuse and aging have degraded. Normal people with intact hippocampal mechanisms simply relearn the forgotten spelling and pronunciation patterns for rarely used words when they are subsequently encountered in everyday life (see Figure 4.1).

However, Henry's hippocampal damage prevented the relearning necessary to offset the degradation of connections that occurs with aging and non-use. As a result, the relatively minor pronunciation, spelling, and word retrieval problems that older adults routinely experience quickly escalated for Henry into major impairments in his ability to read, comprehend, and remember familiar words as he aged.

How to Restore a Memory

Our memories gradually degrade over time if we do not use them. The hippocampus, which governs memory formation, is now thought to engineer the restoration of fading memories in response to experience. For example, if someone has not recently encountered the name of an object such as an abacus, that individual may not be able to recall that name—which had been stored in Broca's area, a region that houses names and labels—when she sees a picture of an abacus. But when she is told the name, the hippocampus swings into action to re-create its memory in Broca's area.

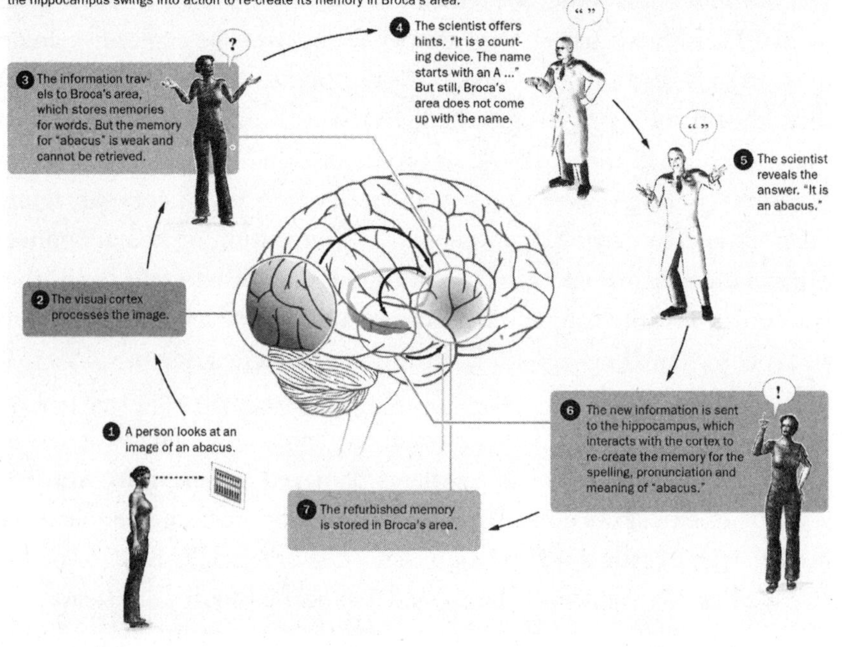

Figure 4.1. How to restore a memory. (Reproduced with permission. © 2014 *Scientific American*, a division of Nature America Inc. All rights reserved.)

But in 1997, Henry's exaggerated aging effects were only a hunch. To determine whether his memories were in fact fading abnormally as he grew older, I needed to document how they changed over time from around age thirty (see Figure 4.2) to age sixty-five (see Figure 4.3) and older.

Figure 4.2. Henry around age thirty-two, five years after his operation. (Photo © Suzanne Corkin, used by permission of the Wylie Agency LLC.)

Figure 4.3. Henry as an older adult around age sixty-five. (Photo © Suzanne Corkin, used by permission of the Wylie Agency LLC.)

My research team first established the trajectory of Henry's memories between age seventy-one and seventy-three. We re-administered some of the word knowledge tests we had given Henry two years earlier. To broaden our window on how Henry's memories changed over time, we next compared his performance on picture-naming and word-reading tests in his seventies versus earlier decades. Finally, we compared data from memory tests that Henry took in his fifties and sixties with his ability to remember similar information in his forties.

AN AGING AMNESIC

By comparing Henry's test scores over time, my lab exposed a dramatic drop-off in his recall of rarely used information from one year to the next—beginning in his forties. Our analyses revealed that Henry's word knowledge didn't just decline as he aged from forty-four to seventy-three years old. This would just show that aging causes decline. Rather, Henry's word memories declined at a *faster-than-normal rate* when compared with memory-normal individuals of the same age. Moreover, Henry's abnormal fall-off with aging only involved rarely used memories. For frequently used words and spelling patterns, neither Henry nor the older adults in our same-age comparison groups showed significant declines relative to young adults eighteen to twenty years of age.[5]

At age fifty-seven, Henry was only slightly worse than normal fifty-seven-year-olds in distinguishing low-frequency words from pseudo-words such as *reversement* in a study by psychologists John Gabrieli, Neil Cohen, and Suzanne Corkin, all then at MIT.[6] However, when my lab gave Henry the same test at age seventy-three, his deficit had grown significantly larger.[7]

Similarly, Henry correctly named pictures on the Boston Naming Test (BNT) without difficulty at age fifty-four,[8] but he had minor difficulty naming the same BNT pictures at age seventy-two in a study by psychologist Elizabeth Kensinger and her colleagues at Harvard University.[9] However, when my lab gave Henry the BNT one year later, at age seventy-three, he made reliably more uncorrected mistakes than normal controls. He also produced new types of uncorrected errors not seen in earlier studies, e.g., neologisms such as naming a *trellis* a "trake," circumlocutions such as labeling *tongs* "ice clippers," and word substitutions, as when he called a picture of a *stethoscope* a "telescope."[10]

Likewise, Henry had a small deficit relative to same-age controls at age sixty-seven when reading low-frequency words in an experiment by Brad Postle, at that time a graduate student of Suzanne Corkin at MIT,[11] but when my lab gave Henry the identical words to read at age seventy-one, he had huge and highly reliable deficits, misreading 67 percent of the words versus an average of 9 percent for our comparison group.[12] Moreover, Henry's problems at age seventy-one escalated when reading the same words at age seventy-three. Over those two years, Henry acquired reliable deficits for various new types of errors not seen at age seventy-one, e.g., affix omissions, as when Henry misread the word *international* as *internal* without correction.

The pattern was similar for spelling: Henry's spelling errors increased dramatically from age seventy-two in Kensinger's study, where he came within one standard deviation of control participants, to age seventy-three in my lab, where he performed more than two standard deviations worse than same-age controls.[13]

Finally, my lab discovered that Henry's slide began in his forties. Our literature search revealed that relative to age-matched controls, Henry at age forty-four remembered the names of actors and politicians he knew before his operation with only mild impairment,[14] but with severe impairment at age fifty-seven.[15]

REFLECTION BOX 4.2: MEMORY'S FOUNTAIN OF YOUTH

Like the rest of the body, the brain tends to deteriorate with age. Yet science suggests ways to lessen the loss.

First, a little perspective. Not all aspects of memory decay with aging. Although slightly slower, older adults are just as able to relearn forgotten information and to understand sentences containing familiar words as they could when they were younger.

In some ways, cognitive function improves with age. For example, vocabulary continues to expand up to age eighty and sometimes beyond. Older individuals score higher on standardized vocabulary tests and spontaneously use a greater variety of words than young people. Conceivably, such age benefits, together with other factors, may contribute to the consistently higher ratings of life satisfaction in the questionnaire responses of older relative to young adults. Only further research can tell.[16]

Older adults do encounter somewhat more trouble when learning new meanings for familiar words and when remembering such things as a telephone number long enough to dial it. Older people also experience frustrating lapses in recalling the spelling of familiar but irregularly spelled words such as *broccoli* and in remembering the names of places and people learned decades earlier.

Recent research, including my own, suggests that older adults can counteract these declines. The key is use of what we know and active exposure to new information. We can prevent deterioration in our areas of expertise—say, public speaking, chess, or playing the piano—by continuing to practice or play. Before meeting up with friends, we can avoid embarrassing tip-of-the-tongue experiences by rehearsing their names in advance. Participating in social situations helps to protect numerous language-related (among other) facets of memory. And we can engage in lifelong learning of various forms. After all, active learning and relearning—reinstating old memories—are ways the hippocampal region keeps us all young.

AGING AND MEMORY MAINTENANCE

Scientists have known for decades that synaptic connections gradually deteriorate with aging, weakening and fragmenting memories that are stored in the neocortex. Seldom used memories suffer the most. If one has not thought about, heard, or seen something recently, that information is vulnerable—and more so, the older one becomes.

The precipitous decline with aging in Henry's memory for rarely used information clarifies how the normal brain helps maintain established memories. The hippocampal complex in older adults must restore degraded memories by replacing them with new memories for the same concepts. Just as a builder can repair damaged structures, so can the hippocampus, in its known role of creating new connections in the cortex, craft fresh memory units to replace broken ones. When normal people re-encounter a forgotten fact in everyday life, they simply relearn it, mending the broken memory and slowing its rate of decline. Without this rebuilding process, older adults would quickly forget the countless fact and event memories they seldom use or try to recall (see Reflection Box 4.3).

REFLECTION BOX 4.3: OLDER BRIDGES, BODIES, BRAINS, AND MEMORIES

Memories and brains share something in common with bridges. For both, two ways to combat effects of aging are available. One way is inexpensive and highly effective, but undervalued. It involves slow, incremental maintenance over decades. The other way is quick and highly valued, but costly. It involves the fast and dramatic process of reconstructing or building anew.

Hundreds of thousands of American highway bridges are currently old and in danger of catastrophic collapse, with predictable loss of life.[17] The good news is that regular maintenance could shore up any bridge for a fraction of the cost of building a new one. The bad news is that money earmarked for ordinary maintenance at every level of government is perennially diverted to new construction projects. Politicians receive praise and votes for building dramatic new bridges but not for maintaining older bridges that fail to collapse.

The story for normal brains and memories is similar. Continuous maintenance is an inexpensive but highly effective way to combat effects of aging. However, people often prefer pricy short-term interventions. The currently popular "brain-training" games and apps are a case in point. Brain-training proponents proclaim claim quick and dramatic real-world benefits from playing these time-consuming and often expensive games. However, a team of renowned memory researchers in the *American Psychological Society* concluded that these brain-training claims are misleading and unsupported by serious science.[18] Indeed, if you reduce your time for social contact and physical exercise in order to play computer games, you might even *impair* your cognitive abilities. Older adults who live alone, exercise little and have few social ties double their risk of developing Alzheimer's Disease.[19]

By contrast, compelling evidence indicates that slow, inexpensive, long-term maintenance of your memories can reduce or delay age-related memory loss and dementia of the Alzheimer's type. *Diet*, *exercise* and *memory use* are the keys to this long-term maintenance process. Many excellent studies have shown that older adults with a *balanced diet* low in fat and cholesterol, and high in vegetables, fruits, nuts, and fish oils perform better on standardized memory tests.[20]

Evidence linking *exercise* to memory function is equally impressive. One longitudinal study measured how quickly older adults could walk thirty feet. The results showed that faster walkers performed better on standardized memory tests over a six-year period.[21] An even more extensive UCLA study revealed a strong correlation between the distance that six thousand older women walked on a daily basis and their risk of developing dementia six to eight years later.

Memory use is the third long-term maintenance strategy for combating memory decline and Alzheimer's Disease. Puzzle solving, education, reading books and magazines, writing, singing, playing a musical instrument, and participating in social activities such as dancing, traveling, volunteering for social causes, and attending concerts, exhibitions, and theater productions are all correlated with better memory and reduced risk of Alzheimer's Disease.[22]

Of course, fact and event memories sometimes become distorted, especially when relearning occurs in a different context from the original learning. The *hindsight bias* illustrates this distortion process. When people are asked to guess the probability of an unlikely event before it happens, and later learn what actually happened, they often forget their previous doubts and now believe they "knew it would happen all along." Knowing the final outcome distorts how they remember their original uncertainty.

ILLUSTRATION BOX 4.4: TWENTY/TWENTY HINDSIGHT

Baruch Fischhoff, a graduate student of Amos Tversky, was a pioneer in showing how *after-the-fact* biases distort event memories. Early in 1972, he asked participants to estimate the likelihood of various possible outcomes of visits to Russia and China that President Richard Nixon had scheduled that year. The possible outcomes included the creation of a joint US-Soviet space program, the arrest of a group of Russian Jews during Nixon's visit to Russia, and face-to-face meetings between Nixon and Chairman Mao in China.

Later, after all three of those possible events took place, Fischhoff asked the same participants to recall their predicted likelihoods for those events. Knowing what actually happened distorted their memory. With high confidence, they recalled assigning higher odds to the events than they in fact estimated—an unconscious hindsight bias.[23]

Tversky viewed this hindsight bias as "a subtle flaw in our reasoning." It leads us to believe that the world is less uncertain than it actually is, and that we are better at predicting events than we actually are.

Overcoming this subtle flaw might seem easy. Simply admit that real world events are inherently complex and unpredictable. Unfortunately, this advice is hard to follow. Overcoming the hindsight bias is difficult. Most people shun ambivalence and doubt. Unconsciously, they like the idea of "knowing it would happen all along." It's comforting.

Distortions in event memories resembling the hindsight bias were never an issue in my studies on the relearning of forgotten words. Distorted relearning of words is unlikely in normal individuals. After people forget a word and then re-encounter it in everyday life, they are motivated to re-acquire and use the word in its standard, undistorted form. Distorted relearning was likewise improbable with Henry, someone unable to relearn *anything*.

Henry's inability to repair and relearn forgotten memories accelerated the rate of decline in his ability to recall rarely used information as he aged. Non-restoration and non-recall made future use of the information less likely. This further degraded his original memories, a downward spiral that allowed aging to erode his neural representations for low frequency words at a faster-than-normal rate. However, Henry did not have to relearn the *common* words that he continued to use throughout his life because frequent use offsets the ravages of aging.[24]

THE REVERSE UNRAVELING ENIGMA

Observations with Henry have often inspired new insights into poorly understood aspects of memory and aging. An example is the curious but well-established fact that amnesics typically remember distant events long after they have forgotten recent events stored in memory during the months and years before the onset of amnesia rendered new learning impossible. This phenomenon seems counterintuitive because normal individuals experience the opposite pattern. In general, we forget distant events before more recent ones. Why should recent memories unravel before distant ones in people unable to learn new information?

The unusually rapid crumbling of Henry's word memories with aging and disuse suggested a simple solution to this "reverse unravelling enigma." To illustrate, imagine that a viral infection destroys the hippocampal complex of a computer programmer named Jenny on her seventieth birthday, suddenly preventing her from learning anything new. Three years after this disaster,[25] a neurologist tests Jenny's abilities to recall two of her memories. One is her memory for the word *motherboard*, a distant memory that she formed at age forty. The other is her memory for the word *touchpad*, a recent memory that she formed at age sixty-nine.

Assuming that Jenny used or encountered each word equally often in daily life before her brain damage (say, once a month), which word will fare better on her neurologist's test: *motherboard* or *touchpad*?

The answer is *motherboard*. With thirty years of use and perhaps also instances of relearning before the onset of amnesia rendered new learning impossible, Jenny's *motherboard* memory will be strong and relatively immune to forgetting during the three-year period between her memory test and when she became amnesic. However, *touchpad* had only one year of use before the onset of amnesia made relearning impossible. This means that her *touchpad* memory will be relatively weak and vulnerable to forgetting during the three post-trauma years before her memory test. By extrapolation, Jenny will experience reverse unraveling. The memories she formed in the years just before the onset of amnesia will fade long before the distant ones that she acquired many years earlier. Beware when familiar memories systematically unravel in reverse, leaving distant memories intact. It may be a sign of the hippocampal damage associated with Alzheimer's Disease.

QUESTIONS FOR REFLECTION: CHAPTER 4

1) What are some advantages of *slowing down* to keep your brain healthy?**
2) What are some ways that you maintain your memories by actively engaging in relearning and reuse of information?**
3) Suppose you have been put in charge of designing a new activity center for people of all ages to maintain their memories. What types of activities would you emphasize in your center?**
4) The holy trinity of long-term cognitive maintenance includes diet, exercise, and memory use. Can you think of any activities you could do to engage more than one of these at the same time to take advantage of their combined effects?**
5) Why didn't Henry's frequently used memories also decline at a faster-than-normal rate relative to normal individuals of the same age?

TEST YOUR MEMORY FOR CHAPTER 4

1) Unlike normal individuals, amnesics remember distant memories long after they have forgotten recent ones formed shortly before amnesia rendered new learning impossible. Why?
2) What is the relation between the reverse unravelling enigma and the extremely rapid deterioration in Henry's word memories as he aged?
3) Henry's brain operation impaired his ability to create memories both for events and for unfamiliar words and phrases. How are the processes for creating episodic and language memories related in the brain?
4) What are three important maintenance strategies for sustained healthy cognitive function?
5) How does the hindsight bias show that memory can be distorted?
6) What are some ways that the hippocampus helps prevent memory loss?

MEMORY TEST ANSWERS: CHAPTER 4

1) pp. 66–67
2) p. 57
3) pp. 65–66
4) Illustration Box 4.3 (pp. 63–64)
5) Illustration Box 4.4 (p. 65)
6) pp. 62–63

Chapter 5
WHAT IS IT LIKE TO BE YOU, HENRY?

FOR HEALTHY COGNITIVE AGING, SHOULD YOU RETIRE OR CONTINUE WORKING?

Many people wanted to know what it was like to be Henry. Henry heard that question in many different forms. However, his answers didn't really help. Someone suggested that reading the transcripts of Henry's interviews was like staring at moving clouds. His nebulous ideas seemed to merge together like clouds.[1] Even when one reader thought Henry made sense, another reader interpreted him differently. Henry spoke a mixed language that seamlessly and unpredictably switched between standard-English and Henry-English, an idiosyncratic language that cannot be translated because nobody else in the world speaks Henry-English. For example, what do you think Henry means by *monograms* in this excerpt from a 1970 interview at MIT? Henry is responding to a request to describe the time just before his operation at age twenty-seven:

> **Henry** ([...] indicates a pause; [??] indicates that a word was inaudible): Before the operation ... and ... always in a way sort of interested in ... uh ... the ... EEG ... uh ... because of ... the ... well, electrical way that it had itself ... and just ... the little things that it could pick up ... and ... it know [??]... and sort of trying to figure it out in your head in a way, just what it ... was able to pick up this and ... just the little things you gave off yourself ... and you ... it was able to follow them ... boy, it knew just when one stopped ... or it faltered ... in fact if it stopped, they naturally

> wouldn't give off . . . but . . . if it faltered . . . you'd see a little difference in the line . . . sort of being . . . monograms there . . . as they were coming through the machine . . .

What Henry means by *monograms* might seem obvious. A sophisticated reader of standard-English might describe the excerpt as a half-spoken metaphor—a beautiful poetic comparison linking monograms to the print-outs from an EEG (electroencephalogram) machine. Technicians interpret the squiggly EEG lines in the print-outs as *brain rhythms* ("the little things you gave off yourself") as easily as if they were reading a monogram, say, someone's initials on a suitcase.

Others might disagree with that interpretation. For scientists familiar with Henry-English, the beauty of that excerpt does not exist in Henry's words but only in the mind of the untutored beholder. Henry did not say that EEG technicians easily interpret squiggly lines as brain rhythms. His words were, "if *it* faltered . . . *you'd* see *a little difference* in the line . . . as *they* were coming through the machine . . ." What do the pronouns *it* and *they* refer to? What is faltering? What "*little difference* in the line" would you see?

Henry's metaphor is equally fuzzy. Henry did not compare anything specific to a monogram. He said that some unspecified (singular or plural) thing "sort of being . . . monograms there." The excerpt is more murky than beautiful.

Worse still, nothing in the excerpt indicates that Henry knew what *monogram* means, either in isolation or as a metaphor. If you asked him, it is virtually certain that Henry could not define the word *monogram*, even if you gave him a literal cue such as, *It can serve to identify a suitcase.* Henry often used uncommon words like *monogram* without being able to define them. Most rare words were incomprehensible to Henry. As you probably do not recall, Henry insisted that the rarely discussed legume known as *lentils* means "area and time of."[2]

It is equally unlikely that Henry could explain what *monogram* means in the metaphoric context *an EEG recording.* Carefully collected experimental data indicate that metaphors were meaningless to Henry.

His ability to comprehend one type of concept in terms of another was severely impaired.[3]

Monolingual speakers cannot translate the uncommon word *monogram* from Henry-English into standard-English without asking "What does *monogram* mean, Henry?" Nor can they imagine what Henry's *EEG monogram* means in standard-English without asking "What is *an EEG monogram*, Henry?" Only my laboratory asked Henry these types of questions. And his answers never made sense.

Was Henry's *monogram* excerpt representative? A news reporter characterized Henry's comments as "compelling" when he described what it's like to *be* Henry.[4] So let's see if Henry became more intelligible when discussing himself. In this excerpt from Marslen-Wilson's 1970 interviews, Henry is answering the question, "Don't you remember things from before your operation quite well?" Do you think Marslen-Wilson was unreasonable to ask, "How do you mean?"

> **Henry**: Yes ... before that, yes [said in a whisper, almost in tears] ... I do remember them ... in a way ... was that ... uh ... was ... the ... worrying part ... and ... always the wondering ... myself of ... just how things are ... and ... well, not just the effect on me but others ...
>
> **William Marslen-Wilson**: How do you mean?
>
> **Henry**: Well ... how it affects me, affects others ... two ways ... my parents, affects them naturally ... either way in a way as you could say ... but the medical profession too ... would be affected because ... they ... would ... uh ... find ... probably that they maybe had done something a little different ... and ... just the correction they could make ... themselves will help others.

Why was Henry *again* vague, incoherent, incomprehensible, and less than "compelling?" One commentator suggested it was because Henry's amnesia prevented him from holding onto ideas long enough to express them.[5] That was not the reason. In his previous comments about monograms, Henry kept the EEG concept in mind for almost a hundred words—more than enough time to express an idea—before Marslen-Wilson gave up and changed the subject.

Henry's core problem was inability to form appropriate new ideas in the first place. In Henry's EEG excerpt, what were the "little things you gave off yourself?" What were "the little things that it could pick up?" What was the "electrical way that it had itself?" What was "*it*?" If asked, Henry would be incapable of answering those questions.

Henry's utterances were like inkblots. Authors interpreted them however they wanted and often quoted Henry's words inaccurately (see Illustration Box 5.1).

ILLUSTRATION BOX 5.1. WHAT HENRY SAYS AS TRANSCRIBED FROM A TAPE RECORDING VERSUS HOW HENRY IS QUOTED

Henry's Response to the Question *What sort of things are you thinking about?* (from the 1970 Marslen-Wilson Transcript)[6]	**Henry's Response to the Same Question as Quoted (in a Nonfiction Book for General Readers)**[7]
Henry: Well ... thought you're wondering just ... the thought yourself, you have is ... it ... how is it going to affect others ... not just what ... way you are you're thinking, but see, is it going to be ... is that the way to do it ... now ... maybe someone else wants it that way, or they want it that way, and ... you have the little ... mental argument I guess you could say, with yourself ... what they ... uh ... what you think ... is ... right that way ... in the mental argument ... and if it ... if you don't put things down directly yourself and say, "hm, that's the way it's got to be done," even if you've thought of all the other ways, that's the way!	**Henry:** "It is a constant effort," Henry said. "You must always wonder how is it going to affect others? Is that the way to do it? Is it the right way?"

Interpretations of Henry's ink blots usually put a positive spin on what Henry actually said. One author saw Henry as a philosopher deep in thought. Another viewed Henry as a poet with a lively sense of humor. Based on transcripts of Henry's tape recorded words, *Remembering* describes scientific analyses that paint a quite different picture of Henry's cognitive and linguistic skills. But first, some background information about Henry.

HENRY'S YOUTH: SUNNY SKIES WITH INTERMITTENT LIGHTNING AND THUNDER

Henry was born in 1926 in a small town about ten miles east of Hartford, Connecticut. His elementary school days were unremarkable. He made friends. He took music lessons.

Things changed at age ten. Henry began having epileptic seizures. One contributing factor was genetic predisposition. Henry's family on his father's side had a history of epilepsy. Another factor may have been a minor head injury at age seven.

At first Henry's seizures were inconsequential. He tuned out for a few seconds, as if daydreaming. Engaged in conversation, he would stop talking, breathe heavily, and then resume what he was saying as if nothing had happened.

These tiny seizures never lasted more than a minute. Even when they occurred daily in junior high, Henry's *petit mal* attacks did not dampen his ambition. He said he wanted to be a doctor or lawyer.

By age fifteen, things deteriorated. Henry suffered *grand mal* seizures. He lost consciousness. His entire body shook. These violent convulsions soon occurred daily and became worrisome. They curbed his outlook. He abandoned his lofty career goals and decided to pursue vocational training in high school.

Nonetheless, Henry's life remained virtually unchanged. His high school education continued unscathed. His teenage social life was normal. He dated. He developed a crush on a much older woman. He held summer jobs. He entertained his friends with popular music on the radio. He built a display case for his collection of pistols and hunting rifles. More than sixty of his high school classmates signed his yearbook, one with the comment, "To a swell fellow and a perfect friend."

Figure 5.1. Henry's high school graduation picture. (Photo © Suzanne Corkin, used by permission of the Wylie Agency LLC.)

After graduating from high school, Henry continued to live at home. His first job was in a junkyard—rewinding the wires in electric motors. This led to a job with Ace Electric Motors, where Henry calculated the voltage in electric circuits and sketched plans for a model railroad. His last major job was to assemble typewriter parts at the Underwood factory in Hartford. Unable to drive, Henry commuted to work with a neighbor but took many days off because of his seizures.

HENRY'S ADULTHOOD: FOG WITH OCCASIONAL STORMS

After his 1953 brain operation, Henry's behavior changed. He became passive, docile, and tractable. He lacked initiative. Henry's caretakers had to remind him to shower, brush his teeth, comb his hair, and turn off the stove or TV. He generally spoke only when spoken to. He would sit quietly for extended periods, neither initiating conversation nor asking questions of the people trying to talk to him. Henry's brain surgery almost certainly contributed to his passiveness. Later studies showed that animals also become passive and tame when the hippocampal region of their brain is removed.[8]

However, Henry sometimes lashed out. Living at home, he fought with his mother. He ordered her to go away when she nagged him. He often became violent. He hit or kicked her. He threatened to kill himself and take her with him to (Catholic) hell. During one heated exchange, he slammed his fist into a door and broke his hand. During another, Henry threw his eyeglasses at his mother.

Love-hate does not fully describe Henry's relationship with his mother. He was curiously indifferent to her feelings. In one strange incident, a caretaker saw Henry's mother lying on the floor "completely out of it." She was not demented but had not bathed or showered for days. Henry was sitting nearby. He could have helped, but seemed completely oblivious to his mother's distress. Henry's emotional coldness to his mother almost certainly reflected in part the damage to his amygdala.[9]

Other people also became the targets of Henry's emotional explosions. His most spectacular outbursts occurred during the twenty-eight years after his mother's death when he lived at the Bickford Health Care Center. There Henry didn't just yell, throw things, and threaten to kill himself. He attacked people so violently that the staff called the police and considered transferring Henry to a mental health institution.

WAS HENRY DEPRESSED? LESSONS FROM A NON-SCIENTIST

According to author and science reporter Luke Dittrich, Henry often felt anxious, worried, and unhappy in 1982. His evidence was Henry's written responses to multiple choice questions in a set of questionnaires called the Beck Depression Inventory (BDI). One BDI questionnaire asked Henry to characterize his feelings over the past few weeks by checking *yes* or *no* against a list of possible mental states. Dittrich inferred that Henry was suffering emotional trauma when he responded *yes* to these dispositions:

"enraged"
"terrified"
"cannot relax"
"frightening thoughts"

Another questionnaire asked Henry to compare his typical feelings with a list of statements. Dittrich concluded that Henry was experiencing high levels of "internal strife" when he checked these statements:

"I feel that the future is hopeless and that things cannot improve."
"I feel that I am a complete failure as a person."
"I am dissatisfied or bored with everything."
"I feel guilty all the time."
"I feel I may be punished."
"I am disappointed in myself."

Was Henry really as emotionally anguished as Dittrich suggested? Dr. Suzanne Corkin had a different opinion. In a published article, she claimed that Henry's responses on the BDI revealed "no evidence of anxiety" or depression.[10]

Perhaps Corkin was right. Maybe Dittrich chose unrepresentative examples. Did Henry check more positive than negative descriptors and statements *overall*? Perhaps Henry did not understand the questions and was responding randomly. I needed Henry's full set of raw data from the BDI to find out.

Professor James DiCarlo, chair of the MIT Brain and Cognitive Sciences Department, had inherited Henry's data files after Dr. Corkin died in 2016. So I wrote Dr. DiCarlo, explaining why I needed a copy of Henry's BDI data. Professor DiCarlo and I had much in common. We were both scientists. We both had strong ties to MIT (where I earned my PhD). We both wanted to vindicate Dr. Suzanne Corkin if possible. After all, Dr. DiCarlo had defended Professor Corkin's reputation as a responsible scientist in a widely circulated publication.[11]

My letter to Dr. DiCarlo followed the guidelines for data-sharing requests endorsed by the American Psychological Association. I was a competent researcher. I would maintain participant confidentiality. I would reimburse any costs that might be incurred. I felt confident that Dr. DiCarlo would send me the data or at least tell me whether they existed at MIT.

I was wrong. Dr. DiCarlo acknowledged receiving my request, but did not otherwise respond. Was it difficult to find Henry's BDI file? This seemed unlikely. I had also requested Henry's 277 filled-in crossword puzzle books and their keys.[12] A file of that size would be easy to find. Did my emails to Dr. DiCarlo get lost or deleted? Unlikely. I had sent duplicate copies and reminders via snail mail.

Perhaps the files I requested had been destroyed. In the digital audio-recording that Luke Dittrich posted on the web, Suzanne Corkin said that she personally shredded many of Henry's MIT data files. Perhaps the ones I requested were among them. That possibility raised serious questions. Why would Dr. Corkin shred those particular files? Dr. DiCarlo had claimed that no significant data were missing from Henry's storeroom at MIT. Now, several years later, why had he not publicly announced the absence of those important files?[13] Only MIT can make those questions go away.

These ethical issues did not interest me. All I wanted was the data. So I asked Luke Dittrich about the BDI responses that he mentioned in his 2016 book.[14] Would he please share Henry's full set of 1982 BDI responses with me? Dittrich did not have to send me those data. I doubted he would. Non-scientists are not obliged to follow scientific protocol.

My mistake. Dittrich immediately emailed me a copy of Henry's entire BDI file.

My analyses indicated that Henry checked negative states such as *nervous* and negative statements such as *I feel that something bad will happen* significantly more frequently than one would expect if he had responded randomly, say, by tossing a coin. Conversely, Henry checked positive descriptors such as *feeling calm* and positive statements such as *I am not particularly discouraged about the future* significantly less often than one would expect if he was responding without comprehension.

I next checked whether Henry's responses were consistent when he took the same test on separate occasions. They were. Henry's data were reliable. Other observations suggested that Henry's data were also valid. His answer sheets contained handwritten notes that qualified several of Henry's responses. For example, after *not* checking *irritable* as a mental state, Henry wrote "maybe a little." He clearly tried to be honest.

Convinced that Henry's BDI responses were meaningful, I calculated his overall BDI score. It fell within the category *borderline clinical depression*—one level worse than the category *mild mood disturbance* and two levels worse than the category *normal ups and downs.* The conclusion was inescapable. Henry's BDI responses indicated that he was often tense, frustrated, anxious, and depressed in 1982. These emotions could easily morph into hostility, which would explain the aggressive outbursts and violent attacks on his mother and fellow residents at Bickford.

Fortunately, the police did not arrest Henry. He never ended up in jail. On the night of his most violent attack, the nurse at Bickford injected Henry with anti-anxiety medicine that calmed him down. Henry's bad behavior also improved over the next few years, according to the staff at Bickford. Henry somehow curbed his aggressive impulses. The upcoming chapter discusses events at MIT that might have contributed to this relatively permanent change and suggests a possible reason why Dr. Suzanne Corkin doubted that Henry's BDI responses indicated genuine depression or generalized anxiety.

QUESTIONS FOR REFLECTION: CHAPTER 5

1) For healthy cognitive aging, what are some advantages of staying positive, keeping a bright outlook, and avoiding negative thinking?
2) What are some pros and cons of retirement versus continuing working on preserving cognitive function as you grow older?**
3) If you had a chance to meet Henry before he died, what are some questions you would ask him to try to better understand what being him was like?**
4) Do you think that Henry-English, as seen in transcripts of Henry's speech, would be more easily understood when talking in person with Henry? Why or why not? Do you think that Henry would have expressed himself more easily in writing instead of speaking? Why or why not?
5) If you were Henry, how would you try to enhance your effectiveness in communicating with others?**

TEST YOUR MEMORY FOR CHAPTER 5

1) What evidence indicates that Henry's problems in expressing himself were not due to an inability to "hold onto ideas long enough"?
2) What was Henry describing when he said "the little things you gave off yourself"? From what sort of device do these "little things" come and what do they represent?
3) Did Henry's seizures during his teenage years severely inhibit his social life, education, and early career opportunities?
4) What does it mean to check the reliability of data from questionnaires or tests (such as Henry's data from the Beck Depression Inventory)?

MEMORY TEST ANSWERS: CHAPTER 5

1) p. 71
2) p. 72
3) pp. 73–74
4) p. 78

Chapter 6
WELCOME TO HENRY'S LIFESAVING CONTRACT WITH SCIENCE!

CAN YOU HELP HENRY HELP OTHERS, YOURSELF INCLUDED?

If Henry was Mr. Hyde at Bickford in the early 1980s, he was Dr. Jekyll at MIT. He behaved himself—the perfect research participant: placid, polite, cooperative, docile, uncomplaining, non-violent.

What accounts for the dramatic contrast between the Bickford Henry versus the MIT Henry? If Henry's angry outbursts at Bickford stemmed from a sense of failure, worthlessness, and hopelessness, why didn't that anger spill over at MIT?[1] It was not just that Bickford reminded Henry of the brain operation that brought his prospects to a halt at the young age of twenty-seven. Every brain scan and experiment on Henry's memory impairments at MIT provided the same reminders.

Nor was it that especially stimulating events at MIT took Henry's mind off his troubles. Henry enjoyed equally distracting activities at Bickford: film nights, bowling parties, Bingo games, and poetry readings. He joined a poker club, a choir, an arts and crafts class, and a Bible study group. He even had pets to play with there. Henry befriended a rabbit, lovebirds, finches, a dog called Sadie, and a singing cockatiel named Luigi. He even had roommates at Bickford and an attractive female friend who held his hand and talked and talked "about everything," including sex. Bickford was no less interesting than his tasks at MIT.[2]

Something else transformed Henry into Dr. Jekyll at MIT. Something more powerful and long-lasting than anti-depressants. Something calming and without side effects. It was a promise that gave meaning and purpose to Henry's life. MIT scientists promised Henry that he would be helping "other people" by participating in their studies of his brain and behavior.

Henry first described his "contract with science" in this excerpt from his 1970 conversations with William Marslen-Wilson. Henry is explaining how he felt about having to answer so many questions in his interviews:

Henry: And . . . I wasn't thinking just . . . uh . . . me giving the answers or asking the questions or anything like that . . . uh . . . I was thinking of other . . . other people.
Marslen-Wilson: Yes . . . how do you mean, "other people"?
Henry: Because possibly it can help you to help others.
Marslen-Wilson: That's . . . that's exactly why we are asking you these questions.
Henry: Uh . . . yes . . . primary thought in a way I guess you could say I have . . . if it helps others, that's **GOOD**, very good.
Marslen-Wilson: It's . . . it's very good of you to look at it like that.
Henry: . . . uh . . . thank you . . .

Henry's contract with science occupied his mind from 1970 onward. He tirelessly reiterated his helping-science-help-others theme to whoever would listen. He announced it on National Public Radio. He boldly injected "helping others" into everyday conversations. Instead of answering questions such as *How do you feel? Where are we? What aspect of remembering are you wondering about?* and *Are you happy?* Henry shifted the topic to helping others. How assertive of Henry! What a contrast with his usual passivity!

By expressing an interest in his memory failures and assuring him that he was helping others, MIT scientists seem to have transformed Henry's feelings about himself from negative to positive. Here are some excerpts from the Marslen-Wilson transcript that highlight this shift in Henry's own words. The first excerpt illustrates the negative emotions Henry initially felt about himself and his memory failures. Note the association in Henry's mind between forgetting and feeling like a low-grade dope:

> **Henry**: Well, when I can't remember things . . . well I just think right off . . . to myself that . . . well, what a dope you are.

In this second excerpt, note that Henry now has positive feelings about how his memory failures are helping others:

> **Henry**: Um . . . I imagine . . . well . . . that always comes at a time . . . that when you want to remember something, that you can't remember it . . . yes . . . well . . . if you can't remember . . . the way [??] I feel anyhow . . . the way . . . I can't remember . . . and it's helping others . . . and that's something right there . . . what I feel . . . myself.

In this third excerpt, note how Henry now sees his memory failures as valuable and interesting to others:

> **Henry**: . . . well . . . whatever it is . . . maybe you don't remember . . . OK, others . . . are interested, they're learning, they're knowing, hearing about it, that's something, right there . . . and . . .—[??] right there . . . figuring that somebody else . . . and . . . somebody else . . .

REFLECTION BOX 6.1. MEANING AND PURPOSE IN LIFE

Anyone can experience feelings of meaninglessness at any stage in life if they feel undervalued, held back, or unable to reach their full potential. Perhaps you're in a dead-end career or a primary relationship that's going nowhere. Maybe you worry that nobody will remember your accomplishments in life after you die. Or perhaps you just never discovered a life purpose during your teenage years. Until people find that meaningful purpose, they often feel empty, as if something vital is missing. They experience a lack of direction, a vague sense of tension, emptiness, dissatisfaction, or boredom, as if caught in circumstances they cannot escape. Occasionally

> these feelings morph into a sense of inferiority, failure, worthlessness, panic, despair and hostility.
>
> Normal people like you and I can easily overcome such feelings. Succeeding in a worthwhile career can satisfy the search for meaning. So can creatively expressing yourself in music, art, or literature. Raising a family and cultivating a capacity for love and intimacy can bring peace of mind. So can overcoming the problems of a dysfunctional childhood. Meeting new people, undertaking new ventures, making new discoveries, travelling to new countries, and volunteering to help others can also add zest and meaning to your life. Paths leading to meaning, direction, and purpose in life are unlimited.

After his surgery, circumstances held Henry in an inescapable trap. Unable to learn anything new, he must have felt permanently held back, lacking in direction, and unable to reach his full potential. His life must have seemed monotonous, meaningless, and without purpose. He probably experienced what teenagers often describe: emptiness, dissatisfaction, and a vague tension, in addition to the feelings of inferiority, failure, and worthlessness portrayed in his excerpts.

Henry lacked the means to overcome such feelings. The usual ways of achieving meaning—love, work, creativity, adventure, and victory over adversity—were unavailable to him. Henry could easily have descended into desperation, panic, and despair. That explains why Henry so enthusiastically described his contract with science as "**GOOD**, very good" (emphasis in the original). Extensive evidence confirms that helping others is an excellent way to counteract negative self-images and feelings of worthlessness and despair.[3]

Henry's contract with science was also important because it gave Henry a way of contributing to humanity at large. No other purpose in life is more comprehensive, meaningful, and significant. By cooperating with his friends at MIT, Henry could "help others" everywhere, now and in the future. Unable to bequeath children, material wealth, or possessions to the world, Henry nevertheless contributed a legacy that will reverberate long after his death.

REFLECTION BOX 6.2: SPARKS THAT LIGHT LIVES UP

Let's say you're like Henry and want to help others. Choose your "others" carefully and keep your effort proportional to the satisfaction you receive in return. If your sole aim is selfless devotion, you will eventually regret it.

For your own happiness and fulfillment, go beyond the big question "What is my purpose in life?" A direct search for meaning could leave you contemplating cosmic significance while doing nothing. Just ask, "How can my knowledge, passions and skills make a difference in the world?" The feeling of contributing is important.

My personal experience in helping others can perhaps illustrate. As a young unmarried man, I wanted to mentor a boy with no father figure in his life. I joined the Big Brothers Organization (now called Big Brothers Big Sisters of America).

My Little Brother was seven when we met. We got together almost every weekend and soon developed a close one-on-one relationship. We had fun. We explored Los Angeles from a child's perspective. We took in movies. We hiked the Santa Monica mountains, rode horses at Hanson dam, surfed at Huntington Beach, sailed in Marina del Rey, kayaked in Newport Harbor, roller skated along Venice Boardwalk.

His mother trusted us. By the time he was fifteen, we spent entire weekends together. I taught him academic, cooking, and carpentry skills. Free of the angst and heavy responsibilities of parenthood, I enjoyed contributing to his intellectual and emotional growth. His grades, confidence, and social skills grew. We celebrated his sixteenth birthday on a three-hundred-mile car and mountain bike trip along the Los Angeles aqueduct.

No longer little, my Little Brother left home to earn his BA in business administration at the University of Pennsylvania. I am proud of this spark that lit my life with purpose and meaning.

You create your purpose in life. The spark that sets your unique world ablaze with meaning depends on your age, your skills, and your passions. Illustration 6.3 illustrates how my life goals shifted with aging. Unlike

Henry, you too can alter your life purpose as circumstances in your life change.

Illustration Box 6.3: PURPOSE AND MEANING LATE IN LIFE

At age seventy, my life required another spark. My entire carpool had retired, and the forty-seven mile commute to UCLA had become intolerable. So I, too, retired and thought about how to help those who had helped me. Recent heart surgery had saved my biological life. Cataract surgery had saved my psychological life. How could I repay such debts to the medical profession? It seemed impossible.

Then I got an idea. I met a student enrolled in a college program just five minutes from my home *by bicycle*. She informed me that students in the program had a BA but needed make-up courses to apply to medical school.

I could not help them with the "killer courses" they were taking in mathematics, physics, chemistry, biochemistry, and organic chemistry. I had barely passed organic chemistry in high school. I nevertheless had skills that could help these potential MDs. I approached the dean of the College with three facts and a proposal. Eleven thousand older Americans (a virtual city) were projected to turn sixty-five every day from 2015 until 2030. To accommodate this demographic onslaught, medical schools were desperately seeking students interested in neurology, gerontology, and psychiatry. Exactly my areas of expertise! I proposed to organize a fun seminar on those topics to help students in the program apply to medical school. The dean enthusiastically endorsed my Brain, Aging, and Mind seminar, or BAM for short.

The students loved BAM and called themselves BAMers. They volunteered to lead seminar sessions. Everyone signed up for an interview project called BAM-B, or Bambi for fun. The goal: to learn from retired physicians about the rewards and challenges of a career in medicine.

I enjoyed helping BAMers gain insights into the unique needs of the older patients they would encounter in their future careers. I relished seeing them develop communication, collaboration, leadership, and thinking skills that would prove useful throughout their personal and professional lives. By helping the medical profession that helped me, I felt fulfilled.

QUESTIONS FOR REFLECTION: CHAPTER 6

1) How can you adopt Henry's strategy of "helping others" in your own life as a way to help preserve the healthy functioning of your mind and brain?**
2) Why was Henry's desire to "help others" after his surgery so easy for him to remember throughout his whole life despite his severe amnesia? What might this tell us about the relation between memory and personal meaning or purpose in life?
3) If you could travel back in time to meet your former childhood self to share some of what you've learned from reading this chapter, what would you say to yourself? What would you share with your relatives or friends?**
4) Henry's newfound passion for helping others by participating in scientific research transformed his demeanor, his interactions with others, and his outlook on life. Considering some problems or difficulties you are facing in your own life, what are some simple ways for you to help yourself by helping others? Do you see a way for a newfound passion to help transform your perceived meaning or purpose in life? Can you remember a time when you successfully did this in the past?**

TEST YOUR MEMORY FOR CHAPTER 6

1) What are some examples in this chapter of how you can retain meaning and purpose in your life? Why is this such an important endeavor, especially for healthy aging?**

2) What does the author recommend is a much better question to ask instead of simply, "What's my purpose in life?"
3) Why was Henry's "contract with science" important for his healthy cognitive aging?
4) Why is it important for medical schools to train their doctors about the psychology of aging?

MEMORY TEST ANSWERS: CHAPTER 6

1) p. 82
2) p. 85
3) pp. 84–85
4) p. 86

SECTION II

THE ENGINES OF MEMORY AND MIND

Chapter 7
THE NEW MEMORY FACTORY

DO ALL ASPECTS OF YOUR MIND DEPEND ON MEMORY?

Research with Henry sparked a revolution in the scientific understanding of memory, mind, and brain that continues to this day. Henry is the main reason scientists now know that hippocampal brain structures play an essential role in forming new memories. Henry also helped the world understand what memories are and why the distinction between new versus old memories is so important.

After William Scoville removed his hippocampus in 1953, Henry's catastrophic memory failures were immediately obvious. He could no longer recall recent experiences. For example, he could not retrace his steps from the hospital bathroom to his bed. He could not say what he ate for breakfast that day. He did not remember where he was or how he got there. Soon after meeting someone, he forgot who they were or whether they had met. He did not know the day of the week or what time it was without looking at a clock. None of this was expected or predicted.

Henry seemed to speak, understand language, and comprehend the visual world normally. Virtually everyone who met Henry saw only one problem: his impaired recall of newly encountered facts and events. His neuropsychologists quickly labeled him an amnesic, a pure memory case. This diagnosis stuck with him for the remaining fifty-six years of his life.

I studied Henry for fifty years (1966–2016). My experimental results confirmed that Henry was indeed a pure memory case. There was a catch, however. Science needed a new definition of memory. Memory encompasses more than just facts and events. In working with Henry, I discovered that thinking new thoughts requires memory. Creating everyday plans requires memory. Detecting speech errors requires memory, regardless of who produces the errors. Perceiving ordinary visual scenes requires memory. Comprehending new ideas requires memory. Imagining new possibilities requires memory. Creativity requires memory.

Difficulty remembering facts and events was not Henry's only problem. His case was more interesting. My results showed that Henry's operation impaired his ability to think new thoughts and to perceive and create the unexpected. He could neither comprehend novel sentences nor coherently and accurately describe fresh experiences. If the definition of "memory" *only* includes facts and events, Henry was not a pure memory case.

REFLECTION BOX 7.1: DO NOT RUSH TO RATIONALIZE THE UNEXPECTED—A LESSON FROM CASE H.M.

Scientists frequently encounter unexpected events that initially seem inexplicable. So do you. Everyone confronts situations in everyday life that at first blush seem mysterious.

What typically happens next is that you and I soon come up with an interpretation, hypothesis, or story that seems to make sense of what happened. And just as quickly, we tend to convince ourselves that our after-the-fact explanation is correct, even when it's not.

The Israeli psychologist, Amos Tversky, observed similar phenomena in experiments on perception at Hebrew University in Jerusalem. Together with Danny Kahneman, a colleague at Hebrew University in Jerusalem and winner of the 2002 Nobel Prize in economics, Tversky concluded that:

"People are very good at detecting patterns and trends, even in random data. In contrast to our skill in inventing scenarios, explanations, and interpretations, our ability to assess their likelihood, or to evaluate them critically, is grossly inadequate. Once we have adopted a particular hypothesis or interpretation, we grossly exaggerate the likelihood of that hypothesis, and find it very difficult to see things any other way."[1]

Reluctance to reconsider initial explanations of unexpected events reflects a systematic bias in how the brain thinks, perceives, and decides. Neuropsychologists examining Henry soon after his operation were as susceptible as anyone else to this built-in cognitive bias. Based on anecdotal observations of his behavior, they quickly and confidently diagnosed Henry as a pure memory case—having no problems besides inability to store new facts and events.

They did not compare Henry's non-memory performance with a normal baseline or comparison group. They did not consider the possibility that there's more to memory than facts and events. They did not contemplate alternative hypotheses that might have changed their mind about their *pure memory* diagnosis. They did not test whether Henry's ability to comprehend visual scenes and to understand and produce sentences was normal. When I took those basic scientific steps decades later, Henry's postoperative language and visual comprehension abilities fell well outside the norm.[2] An earlier generation of researchers had misdiagnosed Henry.

Fortunately, science is a self-correcting enterprise. It is only a matter of time before scientists correct Henry's diagnosis or revise their concept of memory.

OVERCOMING RESISTANCE

My critics initially found my observations troubling. I had difficulty publishing my results. To weaken their resistance, I had to create a new way of conceptualizing mind and brain and compare Henry with a normal baseline in dozens of follow-up experiments. Doing this took many years.

REFLECTION BOX 7.2: HOW TO SAVE A LIFE—LESSONS FROM THE NORMAL BASELINE

People often rationalize unusual events without considering a normal baseline. This story about a young Canadian woman illustrates the potentially disastrous consequences of this basic error.[3] She was in a Toronto hospital undergoing emergency surgery, a victim of a high-speed, head-on automobile crash. X-rays indicated multiple fractures in her ankles, face, feet, and hips.

When fixing her broken bones, her surgeons discovered something alarming. Her heartrate was dangerously fast and irregular, with some heartbeats coming too soon, others not at all.

Her medical team learned that her thyroid gland had been hyperactive for years. They knew that an overactive thyroid can cause heart arrhythmia. They quickly and confidently announced a diagnosis: thyroid malfunction.

At this critical point, her medical team resembled the neuropsychologists who diagnosed Henry a pure memory case. They did not consider baseline norms for people *without* hyperthyroidism. They did not consider alternative hypotheses. They could have killed the woman.

As it happened, she was lucky. Her MDs had ordered thyroid-inhibiting drugs but decided to consult a young skeptic, Don Redelmeier, an internist with statistical training. He noted that heart arrhythmia and hyperthyroidism co-occur very rarely. He suggested a more likely explanation for the patient's arrhythmia: lung failure.

Her doctors reconsidered. They tested Redelmeier's hypothesis and discovered things invisible in her X-rays: hairline rib fractures and a collapsed lung. They ignored her thyroid and treated her lung damage. Her heartbeats returned to normal. She survived almost certain death.

The next day, her MDs checked her thyroid output. It was normal. Her heart and thyroid problems were unrelated.

How did Redelmeier save her life? He asked a basic scientific question: what is the normal baseline—the most common cause of irregular heartbeats? Without leaping to conclusions, Redelmeier considered an alternative hypothesis.

Redelmeier's advice to MDs: when a beautiful hypothesis comes instantly to mind and you're sure it's right, "that's when you need to stop and check your thinking."[4]

Henry's neuropsychologists were sure they were right about the hypothesis that came quickly to their minds after his operation—that Henry was a pure memory case. They needed to stop, check their thinking, and test his non-memory performance against a baseline.

The same advice applies to all of us. Do not quickly explain away unexpected observations in everyday life. Entertain a range of hypotheses then carefully test the validity of each. Especially when someone's life depends on it.

Many of my colleagues were especially reluctant to believe that Henry had language problems. I'll never forget the speaker at a conference who ignored my mountain of solid data on Henry's language errors and announced with a sneer that "Dr. MacKay *thinks* H.M. has language problems."

My reviewers' objections were that language and memory are separate, fundamentally different, and unrelated in function. Memory stores real-world facts, events, and experiences; language communicates facts, events, and experiences to others. Memory is common to all mammals; language is unique to humans. Memory is simple; language is complex.

My critics did concede that the hippocampus did *some* analogous things when creating memories for comparable events and sentences such as experiencing yesterday as beautiful and planning the sentence *Yesterday was beautiful.* To plan that sentence, my hippocampus must conjoin:

- a noun (*yesterday*),
- a verb (*be*),
- a temporal concept (*past)*, and
- an adjective (*beautiful*).

Similarly, to represent my episodic experience that yesterday was beautiful, my hippocampus must conjoin:

- *what* (aspects of yesterday were beautiful), including the many beautiful sounds, sights, and smells experienced during the day,
- *who* experienced the beauty (*me*),
- *where* the beautiful day was experienced (for example, in the mountains), including the trip to and from the mountains,

- and *when* (for example, in the autumn, at age sixty), including the month, day of the week, and year when the day was experienced.

Nonetheless, resistance continued. Decades later, researchers still characterized the hippocampus as unrelated to perception and action—including the comprehension and production of sentences.[5]

DESCARTES'S FRAMEWORK: THE BRAIN AS A TRAIN

A widely held theory proposed in the eighteenth century by the French philosopher René Descartes is the main reason why twenty-first century scientists find it difficult to imagine that storage processes play a role in perception and action, for Henry or anyone else.[6]

Descartes's theory describes the relation between perception, memory, action, and the brain. This was his basic idea. Separate brain modules process information in five or more ordered stages: sensory perception, comprehension-and-thinking, memory-storage, memory-retrieval, and muscle-movement. Sensory-perception and comprehension-thought processes come before memory storage because, in general, information is better retained when understood and contemplated. If needed, action comes next in the sequence, starting with memory retrieval and ending with muscle movement.

I visualize Descartes's sequence as a train with an engine (sensory perception) that pulls a passenger car with a dome (comprehension and thinking), a dining car (memory storage), a baggage car (memory retrieval), and a caboose (muscle movement). Like the cars of a train, the modules in Descartes's sequence are flexible. They can be decoupled or dropped. Retrieval and action need not follow memory storage: we all sometimes act without comprehension or forethought. Descartes's theory also tolerates the insertion of new modules into the sequence, just as passenger trains can add extra cars.

For students of modern psychology and brain science, Descartes's philosophical framework is inescapable. His processing sequence is the organizing principle for virtually all existing cognitive and neuroscience texts. Sensory perception chapters head the index. Next come the chapters on comprehension and thought, followed by memory storage and retrieval. If discussed at all, muscle movement and action occupy the last chapters.

Descartes would love the story about Henry in today's neuroscience texts. The chapter on memory storage says that Henry's operation damaged his module for storing new information but spared his modules for sensory perception, comprehension, thinking, retrieval, and action. It's a wonderful tale.

My critics considered Descartes's framework and the standard H.M. story unassailable. They were not. Giacomo Rizzolatti and a team of Italian physiologists at the University of Parma discovered something that wrecked Descartes's train.

MIRROR NEURONS DERAIL DESCARTES'S TRAIN

The Rizzolatti team located individual neurons in the neocortex that engaged in both perception (for example, seeing someone grasp a grape) and action (for example, reaching out to grasp a grape). They called them *mirror neurons*.[7]

Discovering these perception-action units was accidental. Rizzolatti was searching for *motor neurons*, so called because they connect with neurons that, when activated, directly cause particular parts of the body to move. They were especially interested in motor neurons that moved the *hands*. They wanted to record activity in hand-movement neurons when someone grasped a small object or put it into the mouth. This meant landing tiny electrodes onto individual neurons at precise locations in the brain. Not human brains. They had to settle for the neocortex of macaque monkeys, which closely resembles the human neocortex in structure.

Finding a decent sample of hand-movement neurons in the monkeys required a lot of work. Millions of motor neurons in the macaque neocortex trigger actions in *other* body parts. The team visited dozens of motor neurons before finding a few that erupted with activity when macaques used their hands in various ways.

Aha moments arrived in the laboratory when the computer attached to electrodes in a monkey's brain emitted a loud staccato burst. It signaled activity in a motor neuron. If the same neuron erupted every time the macaque made a particular hand movement, they would announce "It's a grasping neuron," or "it's a mouth-insertion neuron." They named each neuron after its associated movement.

However, the big discovery was accidental. According to one report, a macaque was sitting quietly in its restraining chair, minding its own business, idly waiting for the next task, when it saw the experimenter grab a pencil. *Ratta-tat-tat! Ratta-tat-tat-tat!* The computer suddenly burst into action. This was not supposed to happen. The experimenter dropped the pencil in surprise. Prolonged silence followed. The experimenter picked up the pencil. Again the mysterious burst. The experimenter grabbed other objects. Again the bursts as the monkey watched. Other researchers tried it. Same *ratta-tat-tat!*

This was the *Eureka* moment. Seeing *people* grasp things was activating the *monkey's* motor neuron for grasping. Perception and action in the same neuron!

The team tried other types of actions. When a researcher popped a grape in his mouth, a monkey's motor neuron burst into action. It did not take long to show that this was a mirror neuron for perceiving and producing mouth-insertion.

Rizzolatti's breakthrough discoveries challenged Descartes's framework. For mirror neurons, perception did not happen at one end of Descartes's train and action at the opposite end, with storage, memory, and retrieval cars in between. One and the same neuron was perceiving, representing, and causing an action. Each mirror neuron was a single, indivisible perception-memory-action car. Descartes's imagined separation of perception, memory, and action was wrong.

In 1981, I published a detailed theory of how mirror neurons work in perceiving, representing and producing language and other cognitive skills.[8] Except that my mirror neurons were *hypothetical* neural units. The term *mirror neuron* had not been invented. Real mirror neurons were not discovered for another decade.

In 1998, I showed that my *mirror neuron* theory explained Henry's parallel impairments in perception, action, and memory.[9] By then, mirror neurons had been discovered and I was able to publish my first articles about Henry's deficits in comprehending, storing, and producing new information.

HOW DO HIPPOCAMPAL DRIVERS CREATE LANGUAGE MEMORIES?

To illustrate how the hippocampus creates language memories, let's start with a simple example. I recently learned the name *Sarah MacPherson*. How did I form that memory? I knew some people named *MacPherson* and other people named *Sarah*, so I only had to link those existing word units in my brain to a new or uncommitted cortical unit that could represent the unique conjunction *Sarah + MacPherson*.

My hippocampus played a critical role in linking my neurons representing *Sarah* and *MacPherson* to that new cortical unit. The hippocampus is a type of driver. Like other drivers, hippocampal drivers ramp up the activity in connected neurons in other areas of the brain. However, hippocampal drivers are especially powerful. They cause prolonged and extremely intense activity in connected units, in this case, the old or preformed units in my cortex that represent the words *Sarah* and *MacPherson*. Prolonged activation of those units over a period of seconds—which is an eternity in the brain—created connections to a new cortical unit that represented their conjunction. This is called binding. The same binding process later enabled me to link my new *Sarah MacPherson* unit to semantic information about Sarah that I discovered on the web—her home university, what aspects of brain and memory she works on, her picture, and her excellent book on single-case studies of memory.[10] Binding is how people form new memories and create new ideas.

WHAT ARE MEMORIES?

Memories are clusters of interacting neurons that represent information in the brain. One such cluster represents *Sarah* in my brain, another represents *MacPherson*, and a third supercluster represents *Sarah-MacPherson*. Other neural clusters represent my many memories for ideas, future plans, objects, visual scenes, and musical themes. A cluster of interacting neurons in my cortex even represents the *B* in *Beethoven*.

A few conjoined neurons suffice for representing some memories.[11] Other memories require hundreds of conjoined neurons.[12] The size of a cluster does not matter. All that really matters is whether a memory has a functional internal representation in the brain. Hippocampal drivers form internal representations that are new and functional. Another brain mechanism retrieves internal representations that are already formed and functional. What happens when I initially learn the name *Sarah MacPherson* is fundamentally different from what happens when I retrieve my existing word memory for referring to the many Sarahs I have encountered over my lifetime.

However, functional memories can become defunct or nonfunctional due to aging, non-recent use, and infrequent use over the lifespan. A memory no longer exists when neurons in its internal representation no longer interact or speak to each other. Only if I continue to use my *Sarah MacPherson* memory will it remain retrievable for the rest of my life.

Recently created memories are especially vulnerable. Unless I soon use my newly formed *Sarah MacPherson* memory, the fragile links between *Sarah* and *MacPherson* will become defunct. Her full name would be impossible to recall, but this would not affect my memory for *Sarah* as an independent word. Functional parts of a memory can survive the loss of higher-level links. A chance encounter might even restore the defunct links between *Sarah* and *MacPherson*. My hippocampal drivers could quickly rebuild those broken connections if I happened to glimpse Sarah's name on her book in my office.

YOUR MEMORY FOR A PENNY

Try to draw the two sides of an American penny from memory. Or better still, turn on the record feature of your cell phone and describe as many features of a classical penny as you can. There are at least ten. After you're done, inspect a penny to see how many you remembered.

When I tried this little exercise, I recalled only three features: *round*, *copper-colored*, and the engraved profile of *Abraham Lincoln*. Despite using pennies countless times over my lifetime, I could not recall whether Lincoln's head faced left or right. I did not know where on the coin the date appeared. No recall of the US Treasury building. No memory for the capitalized words *IN GOD WE TRUST*, *ONE CENT*, *LIBERTY*, *UNITED STATES OF AMERICA*, and *E PLURIBUS UNUM.*

Viewing memories as internal representations makes sense of why I recalled so few penny features. I almost certainly did not internally represent all ten features when I first encountered pennies as a child. Unable to read, I lacked the means to create new internal representations for the words *God*, *trust*, *liberty*, *cent*, *United States of America*, or the Latin *e pluribus unum*. My newly formed penny memory probably only contained enough features to distinguish pennies from other coins. The salient visual features *round*, *copper-colored*, and *Lincoln's head* were all I needed.

Those three features received massive use during my life. If I added new features to my penny memory as a teenager, they became defunct due to infrequent and non-recent use over the years.

However, my internal representation for pennies did recently acquire some new and functional features. I needed them to show my UCLA students how my internal representation of an object differs from the real thing and how the adequacy of a memory depends on the task at hand. For example, to accurately draw a penny from memory, my hippocampal drivers had to form many new internal representations, whereas to use pennies in daily life, those formed during childhood sufficed.[13]

My research with Henry further refined the distinction between new versus existing memories. Henry easily read lists of familiar words presented one at a time on cards: *GOT. ATE. STOMACH. HOT. WHO.*

DOGS. ACHES. BOYS. THE. This was unsurprising. Henry's memories for familiar words were intact. He could read familiar words by retrieving functional internal representations formed in grade school.

The surprise came when the experimenter later asked Henry to read the same words in sentences. Henry made spectacular errors and paused abnormally within the sentences. He read *The boys who ate hot dogs got stomach aches* as *The boys* [unusually long pause] *ate hot dogs got stomach aches.* Note the absent *who* in this ungrammatical string, an obvious error that Henry neither noticed nor corrected. Readers cannot make a sentence sound like a sentence by retrieving existing memories for words and phrases. They must create *new* phrases and insert pauses of varying lengths between them, as in *The boys* [short pause] *who ate hot dogs* [major pause] *got stomach aches.*

Henry could not do this. Lacking a hippocampus, he could not internally represent the novel relations between phrases in sentences. This is why Henry failed to insert an appropriate pause to indicate how the subject (*boys*) relates to *getting stomach aches.* This is also why Henry omitted the important little word *who*. He could not see how *boys* relates to *eating hot dogs.* However, Henry deserves thanks for *trying* to read the sentences as sentences. He could have adopted the strategy of reading every word, one at a time, pausing after each, as if reading a list rather than a sentence. This way Henry would have kept the role of the hippocampus in normal reading a secret: He would have made no errors, and without obvious errors to explain, researchers might have missed the significance of his list-like prosody. Fortunately, Henry did not adopt that word-by-word reading strategy.

The upcoming chapter entitled "What's New, Henry?" shows how existing functional memories often allowed Henry to recognize stimuli that intuitively seem novel or never-previously encountered. An example is the fragmented elephant, one of twenty stimuli in a research tool called the Gollin incomplete-figures test. Participants initially see drawings of familiar objects that are so incomplete or fragmented that they cannot say what the objects are. They must guess. Nobody guesses correctly on the first trial. On each subsequent trial they see progressively more complete versions of the same pictures until they can correctly name all twenty objects. When Henry took this incomplete-figures test, he identified as many fragmented objects as normal controls.[14]

Henry had never encountered the fragmented figures prior to his lesion. So why was he so good at this task? "What's New, Henry?" spells out the answer: Henry did not have to form new internal representations of the incomplete pictures *per se*. For example, his existing internal representation of the distinctive tusk, trunk, feet, and tail of an elephant allowed Henry to recognize a fragmented elephant without hippocampal engagement. Intuition is a poor guide to defining what is new versus old in the brain.

MYSTERY SOLVED, RESISTANCE OVERCOME

A principled distinction between new versus old memories also helps explain why neuropsychologists examining Henry shortly after his brain surgery could see only one problem: his inability to recall events. Henry had more difficulty recalling events than facts, not because memory evolved solely to store events, but because we often re-experience facts in everyday life whereas at least one aspect of any event is always new: when it occurs. The now-ness of events is forever new. We can never step into the same cascade of events twice. Henry's brain could not represent the fundamental newness of events in the days, months, and years after his operation.

However, events are not special. Henry's brain registered *no* new information. He could not construct new internal representations of any kind, whether for events, facts, words, or phrases. Asked how old he was, Henry would not recall this fact. He could not create the new internal representations required to update his age on the birthdays that followed his operation. Asked what day it was, Henry would not know because each new day requires a new internal representation. Asked whether he knew Sarah MacPherson just minutes after meeting her, Henry would also say *no* because representing that meeting requires new memory formation.

Memories for salient events that Henry formed before his operation likewise resembled his memories for facts and linguistic information. Henry easily recalled any kind of preformed and functional memory. Asked when he was born, Henry could quickly respond 1926 a fact acquired during childhood. Asked the meaning of common words and phrases that he learned as a child, for example, *hot dogs* and *stomach aches*, Henry could respond without difficulty.

In comparison with his universal inability to recall events, Henry's occasional linguistic lapses in everyday conversation doubtless seemed unremarkable to the early researchers who considered Henry a pure memory case. Henry only produced remarkable errors when forced to create genuinely novel phrases and sentences in the laboratory. Besides, those researchers were not looking for remarkable errors. In their theoretical framework, language deficits would have defined Henry as aphasic rather than amnesic. Later chapters will describe the more complex and interesting truths about Henry that the labels aphasic and amnesic do not express.

QUESTIONS FOR REFLECTION: CHAPTER 7

1) This chapter discusses some of the ways memory permeates our mental lives. Can you think of an aspect of mental life that does *not* depend on some form of memory?**
2) Reflection Box 7.1 discusses the common type of tunnel vision error that occurs when people rush to explain the unknown. What are some ways you could overcome this impulse or to correct your error if you catch it happening?**
3) If brain function is like Descartes's train, what would consciousness be? The conductor, the passengers or the engineer?
4) What are some important ways in which you might or might not be an excellent control participant for establishing a "normal baseline" to compare against Henry in order to better understand how memory works?**

5) Careful scientists usually "entertain a range of hypotheses, then carefully test their validity." Can you think of some situations in your life where it might be important to behave like a careful scientist? Can you think of any situations where it would be impossible to behave like a careful scientist?**
6) Now that you know that the hippocampus integrates or binds together different types of information, how can you use this knowledge to improve your own memory, as when you first learn a person's name?**
7) You have probably seen a computer keyboard countless times in the past. Try to draw one from memory, indicating the exact locations of the letters *B*, *C*, *E*, *I*, *L*, and *Y*. Are you having trouble remembering this information (like drawing a penny in detail)? Now, close your eyes and imagine typing the word *BICYCLE* on an imaginary keyboard. Was this how you previously tried to recall the individual letter positions? Did positioning your fingers in your mind's eye help more letter locations come to mind? What do your observations suggest about different types of memory?**

TEST YOUR MEMORY FOR CHAPTER 7

1) What did Dr. MacKay's research with Henry demonstrate about the role of memory in everyday life? What else is memory for besides remembering facts and events?
2) What is one of the best ways to avoid jumping to premature conclusions, as happened when psychologists labeled Henry a *pure memory* case?
3) Why is it easier for your hippocampus to represent the sentence *Yesterday was beautiful* than to represent the events in that beautiful day?
4) How did discovery of *mirror neurons* that trigger an action and respond to seeing that same action disrupt Descartes's widely accepted framework that viewed the mind as a sequence of

modules that process perception, comprehension, memory storage, memory retrieval, and action?

5) Henry easily read and understood words when presented in isolation but not when those same words appeared in a novel sentence. Can you explain why?
6) If the hippocampus is important for encoding and retrieving information in novel contexts, why could Henry readily recognize familiar objects (and parts of objects) such as a fragmented elephant in the Gollin incomplete-figures test?
7) If someone told you that language and memory are unrelated, what have you learned in this chapter to help them change their mind?**
8) If your brain lacked mirror neurons, what would you be unable to do?
9) How does your hippocampus allow you to learn new names? (Hint: it's a driver for binding.)

MEMORY TEST ANSWERS: CHAPTER 7

1) p. 92
2) p. 93
3) p. 95
4) pp. 97–98
5) pp. 102, 104
6) pp. 102–103
7) pp. 92, 98–100
8) p. 98
9) p. 99

Chapter 8
FETCH THAT MEMORY, BROWSER

CAN YOUR HIPPOCAMPUS HELP PRESERVE YOUR MEMORIES AS YOU AGE?

Chapter 7 examined *how* people form new memories. This chapter examines *why* people form new memories—in order to retrieve and use them. When their brain or the brain of a loved one can't retrieve memories often enough, people become concerned. They want to know why memory retrieval sometimes fails, how they can reduce or prevent such failures in the future, and how the brain normally retrieves memories in the first place.

Retrieval failure is remarkably rare, considering what's involved. Many bilinguals store and retrieve without difficulty over two hundred thousand words in the approximately one hundred billion neurons in their brain. When they need a particular word, say, the high frequency word *memory*, their brain takes only a fraction of a second to activate exactly the right neurons among those hundred billion. How does the brain normally find and activate the internal representation for that particular word so quickly and effortlessly? And on the rare occasions when retrieval fails, how does the brain usually recover the lost word—sooner or later? Research with Henry helped address these questions. It confirmed that the brain retrieves memories in two distinct ways: fast versus slow. Fast retrieval processes were intact in Henry. Slow retrieval processes were not.

FAST RETRIEVAL PROCESSES

Fast retrieval is easy to illustrate. It's what we do every day when we quickly and accurately name familiar objects or retrieve the words in sentences we plan to say. How does the brain achieve this miracle? Speakers themselves do not know. The mechanisms underlying fast retrieval are opaque to introspection. The right words just seem to come to mind and pop out.

Fast retrieval is effortless, unconscious, and requires *browsers*, a type of activation mechanism located in the frontal lobes. Using browsers, we can quickly and effortlessly name familiar objects or retrieve common words in a preplanned sentence during everyday speech. Henry's browsers were undamaged. His frontal lobes were intact. He quickly and effortlessly retrieved the names of familiar objects and the other high-frequency words that he used throughout his life.

The 1890s saw the first breakthrough in our understanding of fast retrieval mechanisms. Two Austrians, the linguist Rudolph Meringer and the neurologist Carl Mayer, observed that speakers only rarely make errors in preplanned utterances, but when errors occur, they are informative. They reveal the basic properties of the brain mechanisms underlying the use of language.

An analogy with the automobile (a new invention at the time) illustrates the guiding concept underlying speech-error research. Just as the occasional backfires in an internal combustion engine can reveal the principles by which automobiles move, the regular ways that everyday utterances break down in speech errors can reveal the principles underlying our ability to rapidly retrieve various types of speech units, including words.

To determine why fast retrieval sometimes breaks down in everyday speech errors, the Austrians carefully collected and analyzed thousands of slips of the tongue. Their analyses showed that speech errors are not random events. They fall into regular patterns that provide an important window on the hidden retrieval machinery in the brain.

SPEECH ERRORS, FRIENDSHIP, AND THE TAPE RECORDER

Meringer and Mayer published their first systematic study of speech errors in 1895 under the (translated) title *Errors in speech and reading: A psychological-linguistic study.*[1] The book contained thousands of German speech errors. Meringer's second monograph contained thousands more errors, organized in the way that eighteenth century biologists might classify different species of plants and butterflies. His procedures set the standard for all subsequent studies of speech errors.

For Meringer, errors were serious. They were important for understanding how the mind works. He carried an error journal everywhere he went. Had he been alive to hear George W. Bush say, "Take the hands out of the guns of people," Meringer would have recorded this in his journal, along with the president's age, his educational level, whether he was reading from a teleprompter (he wasn't), what time of day the error occurred, and what Bush obviously intended, *Take the guns out of the hands of people.*[2]

Accuracy was an important consideration for Meringer. He preferred to collect errors when conversing with friends. This way he could verify what they actually said and ask what they intended to say and what might have caused their error. When relevant, Meringer also recorded words that a speaker had spoken earlier or had recently heard or noticed in print. When conditions were less than optimal for hearing an error or interrogating the speaker, Meringer jotted that down, too.

Completeness was another concern. Meringer tried to record every error he heard. His relentless pursuit of errors apparently disrupted so many conversations that he became unpopular among his colleagues at the University of Vienna. They tended to avoid him.[3] A miniature audiorecorder could have saved some of his friendships.

Ensuring validity was also imperative for Meringer. Because only unintended errors can provide clues to how retrieval mechanisms normally work in everyday speech, Meringer was careful to exclude "intentional" errors from his pool. If *Take the hands out of the guns of people* was fabricated as a fun way for Bush to appeal to voters, it did not belong in

Meringer's collection. Neither did Bush's peculiar pronunciations such as *nucular* instead of *nuclear.* Consistent mispronunciation suggests failure to learn rather than unintended retrieval failure.

The invention of magnetic tape in 1935 added an important new dimension to the rigorous collection procedures that Meringer advocated. The tape recorder enabled carefully controlled studies of speech errors. Many decades later, accurate audio-records became especially important in experiments with Henry. Without physical documentation, some might have rejected Henry's unusual speech errors as incredible.

WORD RETRIEVAL ERRORS FALL INTO PATTERNS

Anticipations are the most common type of word retrieval error. They occur when an about-to-be-produced word replaces an earlier word in a preplanned utterance. An example is *Churches in the . . . I mean, ministers in the church.* The speaker inadvertently activated the upcoming word *church* before its time.

Anticipation errors indicate something important about how the brain retrieves words in sentences. Earlier words do not activate later words in a sentence like bumper cars impacting each other in sequence. For *church* to intrude before its time in the preplanned sequence *ministers in the church*, the speaker must have formed an abstract internal representation of the entire sequence before starting to speak. Some retrieval mechanism (let's call it a brain browser) then activated *church* instead of *minister* in this preplanned internal representation of the speaker's intended output.

Brain browsers are a type of driver that activates stored information. Unlike the general-purpose browser program in your computer, however, there are many brain browsers, each designed to retrieve a specific category of information. Just for words, there are perhaps a hundred different browsers for retrieving word categories such as nouns, pronouns, main verbs, auxiliary verbs, prepositions, adverbs, and adjectives. Hundreds of additional browsers specialize in retrieving different types of facts, events, and images of familiar objects and scenes. Henry could retrieve those preformed memories as readily as normal individuals because the browsers in his frontal lobe were intact.

When a browser retrieves a preformed word memory before its time, the result is an *anticipation error*. *Perseverations* are a related type of error that occurs when a browser revisits an already activated word category in the internal representation of the intended output, as when a speaker intends to say *A project in the future* but inadvertently says *A project in the project*. *Transpositions* are a more complex type of error that occur when a browser retrieves an already-activated word instead of an about-to-be-produced word in the intended sequence, *and vice versa*, as when a speaker intends to say *a computer in our laboratory*, but inadvertently says *a laboratory in our computer.*

As many different categories of units participate in speech errors as there are types of items for sale in a department store. When a speaker wants to say *I always smoke a cigarette with my coffee*, but inadvertently says, *I always smoke my coffee with a cigarette*, the units transposed in error are the noun phrases *a cigarette* and *my coffee.* Here the speaker's internal representation, formed before she began to speak, included both noun phrases and perhaps the entire sentence, *I always smoke a cigarette with my coffee*.

Speech sounds also participate in errors, as when a speaker said *heft lemisphere* instead of *left hemisphere.* Here the transposed speech sounds, /*l*/ and /*h*/, suggest that the speaker's internal representation of her intended output included the full phonology for the phrase *left hemisphere.*

ARE SPEECH ERRORS MEMORY-RETRIEVAL FAILURES?

The brain is not as accurate, reliable, and well-behaved as a computer. Unlike computers, people forget and misremember information. Daniel Schacter, a respected and well-known researcher at Harvard University, made that point memorable in an award-winning book, *The Seven*

Sins of Memory: How the Mind Forgets and Remembers.[4] His book is not about the theology of sin. It sorts different types of memory failure into categorizes that he calls sins of forgetting (or omission), sins of distortion (or commission), and sins of persistence (or perseveration), troublemakers that can disturb peace-of-mind. Schacter's poetic memory sins are intended to help people remember how memory for facts and events fails.

The memory failures that occur during speech production also fit Schacter's rubric. Words commit the same types of retrieval sin as facts and events. Illustration Box 8.1 outlines these correspondences using errors in spoken English from the extensive collection of the late Victoria Fromkin, linguist and dean of the UCLA Graduate Division for many years.[5] These parallels suggest that genuine speech errors are just another type of memory retrieval failure.

Illustration Box 8.1. SPEECH ERRORS AS SINS OF MEMORY

I. Sins of Forgetting

The Sin of Transience. The way experienced events become progressively more difficult to retrieve as time passes is Schacter's sin of transience. The identical sin also occurs when speakers stop mid-sentence, unable to retrieve a rare or not-recently-used word that they know and need. Meringer considered this tip-of-the-tongue phenomenon a type of speech error. William James, the famous American philosopher, considered it a memory failure. It's both. Most speech errors are retrieval failures.[6]

The Sin of Omission. Forgetting aspects of an event is Schacter's stereotypical sin of omission, but speech errors also commit sins of omission. A speaker in Fromkin's collection intended to say *as much AS a surgeon's knife*, but dropped the second *as*, and said, "as much a surgeon's knife," a sin of omission, an error type that is especially common in older adults.

II. Sins of Commission

The Sin of Distortion. The context of recall can distort memories for events, Schacter's sin of distortion. Eyewitnesses to a single-car collision recall a faster rate of impact when asked "How fast was the car going when it smashed into the fence?" rather than *How fast was the car going when it bumped into the fence?* The difference reflects the context of recall, *smashed* versus *bumped.*[7]

Everyday speech errors commit the same sin of distortion. A woman intended to say *left hemisphere*, but inadvertently said *heft hemisphere.* When retrieving */l/*, the context (the upcoming */h/* in *hemisphere*) temporarily distorted her internal representation of the word *left.*

The Sin of Bias. Memory commits Schacter's sin of bias when current thoughts and feelings alter memories of past events. Word retrieval during speech production commits this same sin. Sigmund Freud (Meringer's competitor in the study of speech errors at the University of Vienna) cites a memorable example. A politician wanted to introduce a well-known general as *battle-scarred*, but instead said *bottle-scarred*, unintentionally revealing his covert fear that the general was "hitting the bottle." Attempting to correct this embarrassing error, the lawmaker said, "I mean, battle-SCARED general," inadvertently broadcasting another awkward belief—that the general was a coward. The thoughts and emotions underlying this double-bias sin temporarily distorted the speaker's internal representations for the words *battle* and *scarred.*

The Sin of Misattribution. Memory commits Schacter's sin of misattribution when people inadvertently attribute an internal representation to the wrong source. Unintended plagiarism is an example.[8] Speech errors known as *binding errors* illustrate this same sin. An example from Fromkin is "Rosa always date shranks." The speaker wanted to say, *Rosa always dated shrinks*, but inadvertently linked the concept *past tense* to *shrink* rather than *date*, and said *shrank* rather than *dated.*[9] Henry committed an extraordinary number of binding errors that qualify as sins of misattribution (see Chapter 19).

III. Sins of Persistence

The Sin of Perseveration. Memories of past events often intrude or come involuntarily to mind, a sin of perseveration that can cause insomnia. Speech errors commit this sin in its simplest form. A speaker in Fromkin's collection wanted to say *escorting* but instead said *escorking.* Her already produced */k/* (spelled *C* in *escorting*) inadvertently intruded when she tried to retrieve the subsequent */t/*, a sin of perseveration.

Of course, the repercussions of retrieval sins involving words versus facts versus events can differ in severity. Speech errors are easily corrected sins, unlike distortions of facts and events. Errors in eyewitness recall during a criminal trial can have permanent, deadly, and impossible-to-correct consequences. Nonetheless, the parallels in how word-, fact-, and event-retrieval breaks down suggest that speech errors can offer insights into the hidden mechanisms underlying the retrieval of these other types of memory.

RETRIEVAL CLASSES

Nouns, verbs, and adjectives are *retrieval classes* that govern the sequencing of words in normal everyday speech production. You learned about those categories in grade school, but your teachers probably called them *parts of speech*. Researchers only recently needed the more general term, retrieval classes, to explain why retrieval categories resembling nouns and verbs also govern the sequencing of actions such as lighting and blowing out a match and the sequencing of speech sounds such as the consonants and vowels in *bird* and *word.*

The previous paragraph illustrates some important word retrieval classes: [nouns] (*school, teachers* and *researchers*), [adjectives] (*normal* and *everyday*), [verbs] (*called* and *needed*), [prepositions] (*of* and *in*), [noun phrases] (*your teachers*), and [verb phrases] (*called them parts of speech*). Illustration Box 8.2 provides more systematic examples of words retrieval classes in sentences.

ILLUSTRATION BOX 8.2. RETRIEVAL CLASSES ILLUSTRATED

Article	Noun	Verb	Pronoun	Noun	Preposition	Noun Phrase
The	woman	draped	her	bra	over	the bedpost
The	hiker	erected	his	tent	in	the evening
The	man	tossed	his	briefcase	down	the shaft
Noun Phrase		**Verb**	**Noun Phrase**		**Prepositional Phrase**	
The engineer		missed	her boyfriend		in the evening	
Noun Phrase		**Verb Phrase**				
The guests		hugged their friends at the door				

The retrieval classes [noun], [adjective], and [preposition] govern *word sequencing*: they determine the normal order of words in English phrases, the next level up. Because adjectives precede nouns in English noun phrases, we say *normal everyday speech production* rather than *speech production normal everyday.* Because prepositions precede noun phrases in English prepositional phrases, we say *in everyday speech production* rather than *everyday speech production in.*

The retrieval classes [noun phrase] and [verb phrase] govern *phrase sequencing*: they determine the normal order of phrases in sentences, the next level up. Speakers say *your teachers called them parts of speech* but not *called them parts of speech your teachers* because the normal order for assertive statements in English is [noun phrase] + [verb phrase] (not vice versa).

Teachers probably drilled the labels *noun*, *preposition*, and *verb phrase* into you in grammar school. This deliberate drilling was intended to help you write clear and artful sentences that leapt off the page and into the mind of your readers. Your sentences probably did not do that. Many students found the process pointless, like jumping hurdles when the goal is to harness the imagination using words. They never fully mastered the unnatural circus trick known as writing.[10]

That learn-to-write chapter is behind you. This book focuses primarily on speech. Writing and speech are different. Writing is a human invention that educators explicitly teach in grade school. Not so for speech. Nobody taught you what retrieval categories to learn and use when speaking. You acquired those categories effortlessly and without help when you learned to talk, beginning around age two. For example, you learned the classes [adjective] and [noun] unconsciously and on your own by observing the serial behavior of adjectives and nouns in your native language. Children exposed to English quickly figure out that adjectives precede nouns in English noun phrases.

You didn't even have to learn the strategy of using *categories* to sequence phrases, words and speech sounds. That category sequencing strategy is built into your genes.[11] We know this because children learning a wide range of languages invent their own special categories and sequences of categories that they have never previously encountered. For example, infants with English-speaking parents often create two-word utterances such as *allgone mama*, *allgone dada*, *allgone milk*, or *allgone pacifier*, always with a pivot class, for example, [*allgone*, *bye-bye*], followed by an open class [*mama*, *dada*, *milk*, *pacifier* . . .]. They are already applying a category sequencing strategy to communicate the absence of people and things before they have learned the normal order for noun phrases and verb phrases in English propositions. Their parents might say, *The milk is all gone*, but not *All gone the milk*.

Retrieval classes and the category sequencing strategy evolved in the human genome over millennia to solve several problems. One was how to efficiently create complex sequences of internal representations that can be transferred from one human brain to another.[12] Binding mechanisms in the hippocampus were the key player in solving that evolutionary problem.

Another problem was how to quickly retrieve and use internal representations when needed during everyday speech and action. Browsers in the frontal lobes were the key player in solving that evolutionary problem. Browsers quickly activate specific categories of phrase, word, or phonological units. The substitution regularity in speech errors shows that.

THE SUBSTITUTION REGULARITY IN SPEECH ERRORS

Substitutions are the most common type of retrieval error in English, German, Dutch, and many other languages.[13] They occur when one unit inadvertently replaces another in an intended utterance. These replacement errors are interesting because they are not random. A reliable statistical pattern known as the *substitution regularity* describes substitutions in a wide range of languages: when speakers inadvertently substitute one unit for another in error, the intended and substituted units virtually always belong to the same retrieval class.[14] I'll illustrate this regularity with examples of phrase, word, and speech sounds in Fromkin's collection.[15] If I translate this book for publication in German, I could easily illustrate the substitution regularity using German examples from Meringer's collection.[16]

Phrases are the largest units substituted in error. An example is, "I have to smoke my coffee with a cigarette," where the speaker intended to say *I have to smoke a cigarette with my coffee* but inadvertently substituted one noun phrase (*my coffee*) for another (*a cigarette*). Like thousands of other substitution errors, this phrase transposition obeys the substitution regularity. Noun phrases *always* substituted with other noun phrases in Fromkin's collection.

It is not that violations of the phrase substitution regularity are impossible to imagine or to produce. One can easily envisage a hypothetical speaker intending to say *I have to smoke a cigarette with my coffee* who inadvertently substitutes the noun phrase *my coffee* for the compound verb *have to smoke*, yielding the error *I my coffee a cigarette with have to smoke*. However, real substitution errors are not like that. Cases where a normal speaker inadvertently substitutes a noun phrase for a verb phrase have never been reported. Verb phrases in Fromkin's collection always substituted with other verb phrases, as when a speaker said, "If you meet him you'll stick around," substituting one verb phrase (*meet him*) for another (*stick around*) in the intended utterance *If you stick around you'll meet him*.

Individual words also obey the substitution regularity, as when a speaker intended to say *the native values* but inadvertently said *the native vowels*, substituting one noun (*vowels*) for another (*values*). Nouns always substituted with other nouns in Fromkin's collection. If instead the speaker had said *the native restless*, this hypothetical substitution of an adjective (*restless*) for a noun (*values*) would have violated the substitution regularity. Normal speakers don't do that. In available error collections, nouns substitute with other nouns, not with determiners, verbs, auxiliary verbs, prepositions, or adjectives.

Speech sounds also obey a substitution regularity, as when a speaker in Fromkin's collection inadvertently said *taddle tennis* instead of *paddle tennis*, substituting the syllable-initial */t/* in *tennis* with the syllable-initial */p/* in *paddle.* Syllable-initial consonants always replace other syllable-initial consonants in error; they never substitute with vowels or syllable-final consonants. Again, violations of this regularity are easy to imagine and produce. A hypothetical speaker could easily say *saddle tennis* instead of *paddle tennis*, substituting the syllable-final */s/* in *tennis* with the syllable-initial */p/* in *paddle*. However, normal speakers do not do that. Syllable-final consonants always substitute in error with other syllable-final consonants, as when Fromkin heard a speaker say *wish a brush* instead of *with a brush*, inadvertently substituting the syllable-final sounds in *brush* and *with*.

RETRIEVAL CLASSES, BROWSERS, AND THE BRAIN

Research with aphasic patients provides important clues to the locus of category-specific retrieval mechanisms in the brain. Aphasic patients with damage to left prefrontal brain regions don't just have difficulty retrieving individual words or phrases. They fail to retrieve entire *categories* of words and phrases. This phenomenon, known as category-specific aphasia, suggests that the browsers for retrieving *categories* of phrases, words, and speech sounds reside in the frontal lobes.

Patients I'll call Mary, George, and Betty are category-specific aphasics. All three have left prefrontal damage. On standard tests of aphasia, they each omit and transpose entire categories of words and phrases, although the impaired categories differ from one patient to the next.[17]

Mary has problems with pronouns, determiners, and auxiliary verbs, as became obvious when she described the cookie theft picture, a standard test stimulus. Normal speakers describe the picture in coherent sentences like these:

A woman is standing in her kitchen doing dishes. She does not see the water flowing out of the sink onto the floor in front of her. She does not notice the boy and girl behind her. The boy wants to give the girl a cookie. He is reaching into a cookie jar high up on a shelf. The stool he is standing on is about to fall over.

Mary's description was different: "Cookie jar . . . fall over . . . chair . . . water . . . empty . . . ov . . . ov . . ." [18]

Note that Mary said "fall over . . . chair" with the verb preceding the noun. This violates the standard English noun-before-verb word order seen in *The chair is about to fall over.* Mary's description of the cookie theft also lacked coherence, as if she was simply labeling isolated aspects of the picture. Mary's description was also impoverished. She used verbs (*fall over*) and nouns (*jar*, *stool*, and *water*), but no pronouns (*she*, *her*), determiners (*a*, *the*), or auxiliary verbs (e.g., *is*, as in *is reaching* and *is standing*). It is as if Mary's browsers for activating the specific categories [pronoun], [determiner], and [auxiliary verb] have been damaged but the ones for activating the categories [noun] and [verb] have been spared.

Other patterns emerge when different category-specific aphasics describe the cookie theft picture. Some omit nouns, as if their browser for activating the [noun] category is impaired. Others omit verbs, as if only their browser for activating the [verb] category is dysfunctional.[19]

A more complex pattern surfaced when George heard and then retold classical stories, another standard test for aphasia. This is roughly how normal speakers retell the fable of the fox and crow:

The fox uses flattery to trick the crow into opening its mouth and dropping its cheese, which the fox then steals.

George retold the fox and crow fable quite differently:

"Well . . . well . . . the same thing is s-smart everything, smart . . . and the brain, OK."

Here George used nouns (*thing*, *brain*), adjectives (*smart*, *same*), and articles (*the*), but no pronouns (*its*), infinitives (*to trick*, *to steal*), or verb participles (*opening*, *dropping*). It is as if George's browsers for activating the categories [pronoun], [infinitive], and [verb participle] are damaged.

Betty illustrates another unique pattern. This is her retelling of the fox and crow fable:

"King . . . Singing . . . Singing loud . . . Meat. Perfect!"[20]

Betty included nouns (*king*, *meat*), adjectives (*loud*, *perfect*), and present participles (*singing*, *singing*), but no main verbs, an essential category in grammatical sentences. It was as if Betty's browser for activating the category [main verb] is impaired.

Other category-specific aphasics have a related problem involving speech sounds: they omit, substitute, and transpose entire categories of phonological components in common words that they retrieved without difficulty before the onset of aphasia. They consistently produce jargon words such as *marmer* instead of *barber*, *supei* instead of *toothbrush*, *deks* instead of *desk*, or *ragon* instead of *wagon*. Again, however, the pattern varies. Each patient seems to end up with a virtually unique jargon vocabulary. It is as if different phonological browsers are impaired in different patients.[21]

SUBSTITUTION ERRORS IN PROBLEM SOLVING: A DIFFERENT FETTLE OF KISH

Daniel Kahneman, a psychologist and winner of the 2004 Nobel Prize in economics, discovered a type of substitution error that occurs during problem solving and thinking, a level well above speech sounds, words, and phrases.[22] Kahneman and Shane Frederick, a colleague at Princeton University, asked people to solve moderately difficult problems and found that they often substituted and solved an easier problem without awareness of their error. To illustrate this high-level substitution phenomenon, imagine that an experimenter asks you to think about your love life and then asks, *How happy are you?* Chances are you will not answer that question. Instead you will substitute and answer an easier question: *How happy is your love life?* Unwittingly, you have swapped problems.

As another illustration, try solving this problem. A bat and a ball together cost $1.10. The bat costs $0.10 more than the ball. How much does the ball cost?

If you are like most people, you unconsciously replaced that challenging problem with an easier one. You broke the sum, $1.10, into two easy-to-think-about parts, $1.00 and $0.10, and responded $0.10. That answer is *incorrect*. To arrive at the *correct* answer ($0.50), you needed to apply some algebra. If you did not, don't feel badly. Very few *statisticians* presented with this bat and ball problem solved for BALL in the equations, BAT + BALL = $1.10 and BAT - BALL = $0.10.

Now try answering this question. Are Kahneman's thought substitutions like substitution errors in speech? My answer is NO. Although they share the same *label*, substitution errors in thought and speech are fundamentally different.

Substitution errors in speech are memory *retrieval failures*. They occur when speakers retrieve units in the internal representation of what they intended to say before beginning to speak. This makes speech errors easy to correct. By revisiting the existing internal representation of their intended output, normal speakers can quickly replace what they said in error with what they intended to say.

By contrast, Kahneman's thought substitutions are memory *formation* errors. They occur when a thinker inaccurately represents a newly encountered problem and solves a different problem. This makes thought substitutions *difficult* to correct. After making a problem substitution, thinkers must go back and rebuild an accurate internal representation of the original problem.

The principles that underlie substitutions in speech and thought also differ. A same-class substitution principle characterizes speech errors (substitute one member of a retrieval class for another in the same class), whereas a laziness principle characterizes thought substitutions (substitute an easy problem for a hard one).

To easily remember these contrasting substitution principles, imagine two professional baseball coaches, Coach Lazy and Coach Class. Both follow less-than-optimal strategies when substituting pinch-hitters into their batting line-ups in the final innings of a game. For his team, Coach Lazy consistently selects pinch-hitters based on their readily available seasonal batting averages, without considering their harder-to-determine batting success against the particular pitcher on the mound (an important factor). Coach Lazy asks the pinch-hitter with the highest seasonal average to go first, then the one with next highest seasonal average, and so on.

By contrast, Coach Class reliably substitutes pinch-hitters based on their job category: he asks a pitcher to pinch-hit for a pitcher, a catcher to pinch-hit for a catcher, a third baseman to pinch-hit for a third baseman, and so on.

Coach Lazy is following the laziness principle seen when problem solvers unconsciously switch problems. Coach Class is following the same-class principle seen when speakers unconsciously substitute one speech unit for another. There probably are baseball coaches like Coach Lazy, but almost certainly none like Coach Class. When you think about it, retrieving words from the brain is a rather odd game, quite unlike baseball.

SLOW RETRIEVAL

The *fast* retrieval processes discussed up to now differ from *slow* retrieval processes. Slow retrieval involves conscious and effortful processes that resemble problem solving. It's a back-up strategy that normal individuals often adopt to retrieve hopelessly lost words. Curiously, however, Henry did not seem to use this back-up mechanism to resolve his memory-retrieval failures. Slow retrieval seems to require the binding mechanisms that Henry's brain operation destroyed.

Slow retrieval becomes necessary when fast retrieval fails despite repeated attempts. It is an active approach to combating memory loss that can resemble problem solving. As with fast retrieval, already formed categories guide slow retrieval. However, speakers create their slow-retrieval categories anew after retrieval fails, whereas standard, preformed categories such as [proper name], [adjective], or [verb] guide the normally error-free process of fast retrieval.

For example, Raymond Nickerson, a memory researcher at Tufts University, concatenated the standard preformed categories [friend], [female], and [first name] in order to create the slow-retrieval category [first name of a female friend] after the fast-retrieval process failed to retrieve the name of a familiar street near his home. Nickerson wrote, "The name would not come to mind, but I did know it to be the name of a friend. The name *Elliot* suggested itself, but did not seem to be correct. I thought the name I was looking for was a first name, and although *Elliot* can be either a first or last name, it is in fact the last name of a friend of mine. I was also fairly sure the sought-for name was the first name of a female, and the Elliot in mind was a male . . ."

"As the search continued, the name *Cellier* surfaced, the last name of a close friend of Elliot's, who was also a friend of mine. Next came *Emil*, the first name of Cellier; then Hilda, wife of Emil." Nickerson immediately recognized Hilda as the name of the street.

Note how Nickerson retrieved and rejected a chain of associated names before his forgotten street name *Hilda* finally popped up. For example, Nickerson had to reject the last name of his friend *Elliot* because it did not fit the provisional category he had just created: [first name of a female friend]. Henry solved none of his many memory problems in that way. If he queried his own memory, he did not pursue what came up. To see why, let's take another example.

In this case, the newly created categories for guiding slow retrieval were more specific and complex: [the name of an amusing singer, two PhDs in ecological psychology, met briefly at an event three years earlier]. Using those three categories, a friend of mine in his mid-seventies recalled the name of someone he met briefly at an event three years earlier. Browser activation only dredged up two aspects of this hopelessly lost memory: the singer was especially amusing and had two PhDs in ecological psychology.

Unable to recall even a single speech sound in the singer's name, my friend used his slow retrieval category to query his memory into the night. Initial search results were vague and tentative, then sharpened. The singer's missing name felt somewhat Jewish, then *very* Jewish. Finally a breakthrough. The name *Cohen* burst into awareness!

Confident that *Cohen* was the correct last name, my friend went online and Googled *Cohen ecological psychology*. This gave him *Michael Cohen*, head of Project NatureConnect—an institution that awards PhD degrees!

To ensure that *Michael* Cohen was the right person, my friend created another new search category [singer's amusing song]. Again, slow retrieval progressed from nebulous to precise. The song felt "folksy." Was it about dancing? Fragments of a familiar lyric came haltingly to mind: "Buffalo gals won't you come out tonight and dance by the light of the moon." An image of a girl with a hole in her stocking sharpened into focus.

He again queried the Web. "Hole in her stocking" yielded nothing, so he Googled "Buffalo gals" and up came—

"Buffalo gals won't you come out tonight
And we'll dance by the light of the moon."

Eureka! This was definitely the singer's chorus. But the verses felt wrong. What was missing? The gal dancing in his memory was somehow *amusing* and *endearing*. Creating the novel search category [amusing, endearing female dancer] revealed that it was not her *clothes*. *Hair* triggered no associations. Was it her *face*? This query triggered another memory fragment: "danced with a gal with *X* on her face!"

What concept occupied the *X* category? It wasn't dirt, wrinkles, or pimples. The image sharpened. *Freckles*! She had freckles on her face. Entering *Michael Cohen freckles* into his search engine, up came:

"I danced with a gal with freckles on her face
Freckles on her face, freckles on her face
I danced with a gal with freckles on her face
And we danced by the light of the moon.
When I asked her where she got 'em, she said,
"I've got 'em all over, Got 'em all over,"
When I asked her where she got 'em, she said,

"I've got 'em all over, Got 'em all over,"
And we danced by the light of the moon."

Problem solved. The singer's name was *Michael Cohen*.

How did my friend retrieve this hopelessly lost information? He created novel categories for actively quizzing his memory, paying close attention to the vague and fragmentary answers that came up. He then sought help from his computer, a machine thousands of times faster than neurons that, unlike people, never forgets unused information. Combining input from mind and machine, my friend pieced together his memory fragments into a picture, as if solving a jigsaw puzzle.

Most Internet users can achieve similar feats of slow retrieval in collaboration with a computer. All it takes is imaginative use of Web skills—which will not decay with aging if you continue to use them. The same is true of other skills. Continue to practice the skills you enjoy—say, carpentry or playing bridge.[23] Together with exposure, use, and relearning (including computer-assisted relearning), this active approach will combat memory loss.

REFLECTION BOX 8.3. USE YOUR HIPPOCAMPUS TO LOSE WEIGHT

Are you interested in losing weight? Do what Henry could not do. Let your hippocampus help! If the sights, smells, and sounds associated with food tempt you to reach for snacks and seconds that your body does not need, engage your hippocampus and your imagination in creating new internal representations of food that you recently ate and are about to eat. Begin by imagining how enjoyable your last meal tasted. Then imagine your next meal—what you will eat, with whom, when, and at what restaurant. Surprising new research shows that this little exercise can help you eat less and snack less between meals![24] The simplest explanation is that creating new internal representations of your recent and upcoming meals can call up a feeling of being full. This could help you trim down by making you less interested in eating more.[25]

Now for a caveat. Exposure, use, relearning, and practice only work to facilitate memory retrieval within narrow limits. They will not restore memories *across the board.* Using and relearning a particular word might fortify your memory for that word and perhaps other words with similar spelling or pronunciation, but won't help maintain the millions of other memories stored in your brain. Use will strengthen a *particular* skill, but the *overall* rate of cognitive decline *across the board* does not differ for people who practice solving crossword puzzles versus those who do not.[26]

EXERCISE: THE MIRACLE MEDICINE

The only proven way to preserve *overall* brain functioning as you age is physical exercise. It's the most powerful component of the active approach to memory maintenance. Recent research shows that forty-five minutes of moderate aerobic exercise three days a week can build healthier and stronger neurons in the brain and increase the volume of the hippocampus by 2 percent (a big accomplishment because the volume of the hippocampus declines by about one percent a year after age fifty).[27] Brisk movement can also help you focus better, remember better and improve your mental health. It can enhance your energy level. It may even reduce your risk of acquiring Alzheimer's symptoms as you age.[28]

Exercise can also tune up the remaining thirty-five-trillion cells in your body. This can help you feel better, look better, and sleep better. Your heart, circulation, digestion, and immune systems may work better. Exercise is beyond doubt a miracle medicine. And it can be free![29]

Combining moderate exercise with meaningful memory use, learning, and relearning is by far the most efficient way to maintain your memories as you age. As an older adult, I walk my talk. I routinely strengthen body and brain by combining physical exercise with memory use. My dogs and I enjoy long walks with friends in state and national parks near my house. The many topics we discuss keep particular memories sharp, and the vigorous hiking revitalizes memories across the board. Talking while walking has also enhanced my friendships and social support. One more benefit: my dogs love me more with every hike.

I also enjoy the mental and physical sport of aqua-aerobics. I regularly join a small and congenial group of like-minded colleagues wearing flotation belts that keep our heads above water as we exercise in a deep pool. Our noon-time meetings simultaneously combine zero-impact activity with social communication and active learning in ways known to offset declines in a wide range of cognitive and physical abilities. We can work our muscles as vigorously as we like in continually changing routines that seem effortless because we are discussing an endless variety of topics, including personal issues, our fields of expertise, recent films, colloquia, concerts, sports events, campus gossip, and especially jokes.

REJOICE IN GROWING OLDER

Robert Browning encouraged us to rejoice in growing older: "The best is yet to be, the last of life, for which the first was made." Sadly, the last years of Henry's life were not "the best." He maintained neither body nor brain as he grew older. The healthy rainbows of action and experience enjoyed by my hiking and aqua-aerobics groups? Unavailable to Henry. Coherent contributions to our discussions? Impossible for Henry (see Chapter 19). Our jokes? Incomprehensible to Henry (see Chapter 23).

Henry could not even reap the benefits of exercise in his later years. Drugs prescribed since 1953 for controlling epilepsy eventually damaged his cerebellum, which reduced his mobility. Further impeding movement, Henry became seriously obese. Unable to remember his last meal or imagine his upcoming meal, Henry overate. He especially overindulged his love of the high-calorie foods offered by his keepers.[30] Generous to the end, Henry bequeathed his brain to Dr. Suzanne Corkin, despite being unable to recall who she was. Confined to a wheelchair, he died of congestive heart failure in 2008.

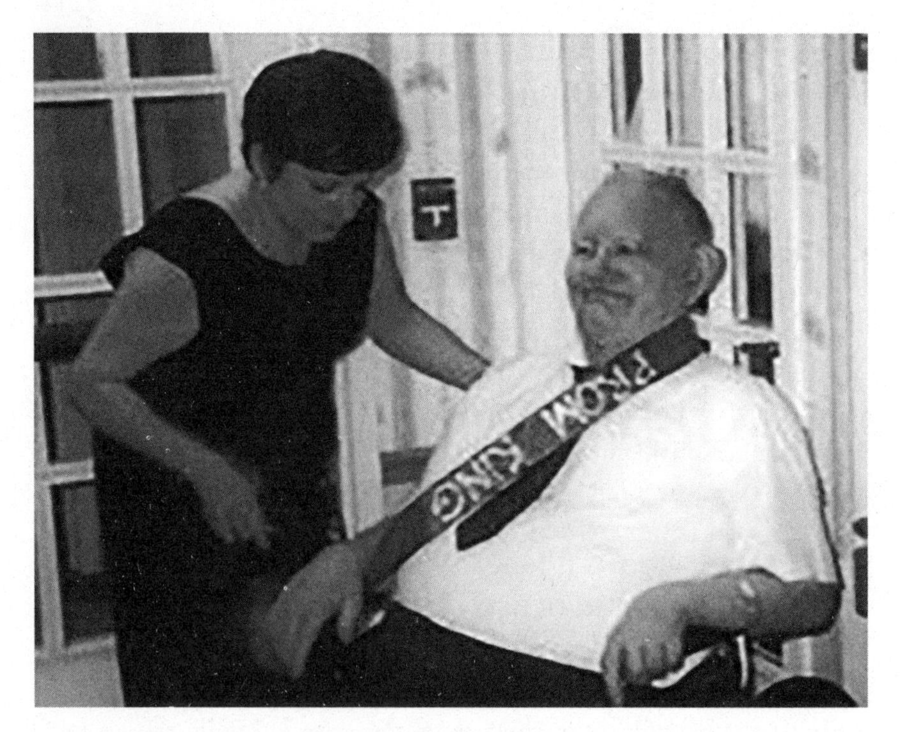

Figure 8.1. Henry late in life. (Photo © Suzanne Corkin, used by permission of the Wylie Agency LLC.)

QUESTIONS FOR REFLECTION: CHAPTER 8

1) Compare and contrast a brain browser versus an Internet browser. What would an Internet advertisement correspond to in a brain browser? How about the "back button" or bookmarks?
2) Which does the human brain need to learn more easily: speaking or writing? Why?
3) Recall the role of the statistical pattern called the substitution regularity in understanding speech errors. Can you think of examples where the substitution regularity might apply in other cognitive domains, such as musical performance, mental mathematics, social interactions, or performing complex actions?
4) Have you ever known or encountered someone, perhaps in line

at the airport, whose speech resembled category-specific aphasia (resembling the speech of Mary, George, or Betty described earlier in this chapter)? Could you identify which retrieval classes were missing in their speech? Which absent retrieval class would disrupt your understanding the most?**

5) This chapter ends with a discussion of the author's lunchtime aqua-aerobic group. What other ways would you find enjoyable to combine exercise with learning and relearning in order to better maintain your memories as you age?**
6) This chapter discusses computer-assisted relearning, as when you search online for a name that you temporarily forget. Is it possible to use computer-assisted relearning too often? Do you think that today's tech-savvy and tech-dependent youth will experience more or fewer memory problems in old age relative to earlier generations?**
7) What are some ways that your brain is like a muscle? What are some ways that it's not?

TEST YOUR MEMORY FOR CHAPTER 8

1) What are some ways you could combine exercise with learning and relearning to better maintain your memories as you age?**
2) Which general area, or lobe, of the brain is most likely to contain the browsers for fast retrieval? Why did Henry have no difficulty with fast retrieval?
3) What is the cognitive scientist Rudolph Meringer most famous for?
4) What evolutionary advantages did the development of retrieval classes and category sequencing bestow on the mind and brain?
5) Why is the substitution regularity important for understanding speech errors?
6) What is the main difference between substitution errors in speech versus thinking?

7) What is the main difference between fast versus slow retrieval? What brain mechanisms underlie this difference?
8) How and why does exercise benefit brain function?
9) When fast, automatic retrieval fails, as when you cannot recall the name of someone you know well, how can the alternative, the slow, deliberate, and active retrieval process, help you overcome that failure and memory loss in general?**

MEMORY TEST ANSWERS: CHAPTER 8

1) p. 126
2) p. 108
3) p. 108
4) p. 116
5) p. 116
6) pp. 120–21
7) pp. 122–23
8) pp. 126–27
9) pp. 122–23

Chapter 9

HUMPTY DUMPTY AFTER HIS FALL

IS YOUR BRAIN BUILT TO FORGET?

For Plato, memory was simple. The core of the brain contained a wax-like substance that represented memories. Just as a key pressed into ordinary wax leaves an impression that one can use later to re-create the key, experiences left impressions on this wax-like matter that the brain could later use to recall or reconstruct the original experience. The experiences could be events, facts, feelings, actions, concepts, or images. It did not matter. Memory-wax anywhere in the brain core could register any type of information.[1]

For Plato, the *process* of leaving an imprint on the memory-wax was also the same throughout the brain. However, the quality of the wax could vary between individuals. People with better memories had more flexible memory-wax. Experiences registered deeper and sharper impressions in their brains.

At the time of Henry's operation, Plato's basic idea dominated the field of memory. The prevailing view was that the same process stored any type of information anywhere in the brain. Searching for brain areas dedicated to forming specific types of memory was pointless. Specific sites in the brain core might create and store different types of memories, but these specific regions were thought to vary from person to person.

Research with Henry radically altered that view.

HOW HENRY SPLIT HUMPTY DUMPTY IN TWO

In 1955, Henry sat at a table in an office at Hartford Hospital. Before him was Professor Brenda Milner, a renowned psychologist who had traveled from Montreal to meet him. A pencil in Henry's right hand rested at the apex of a large five-pointed star drawn on a sheet of paper. The star had a double outline, with a half-inch corridor separating its inner and outer borders.

Dr. Milner asked Henry to trace counterclockwise around the star, keeping his pencil inside the corridor. This was difficult because Henry could only see the star in a mirror that reversed up and down. A barrier prevented direct sight of his hand.

When Henry began, Milner clicked a timer that she stopped when Henry returned to his starting point on the star. She recorded Henry's time, replaced his traced star with a fresh one, and asked Henry to begin again. He traced the identical star fourteen times a day for three days. Who would have guessed that Henry's behavior in this task would tear asunder what everyone had believed about brain and memory for millennia?

By trial fourteen on the first day, Henry's performance had improved. He strayed outside the corridor less often and circumnavigated the star faster than on trial one. Henry could learn something new after all!

On day two, Milner showed Henry the same star and asked if he had seen it before. Henry shook his head. "Nope." He did not remember the fourteen trials on the previous day.

Milner redescribed the task and Henry did it again … almost as adeptly as on the last trial of the day before. Milner vividly remembers Henry remarking, "Well this is strange. I thought it would be difficult, but it seems as though I've done it quite well."[2]

For Milner, this was exciting. Henry's inability to recall the earlier test-taking event made sense. The previous day's test was an event, and Henry could not remember recent events.

But memories for movements must be different. Henry could recall *how* to do something from one day to the next without remembering the *fact* that he had done it before. Repeated trials had *slowly* improved Henry's tracing ability without his awareness.

Henry's ability to learn muscle movement procedures but not facts seemed revolutionary. It suggested that Plato's widely accepted view must be wrong. The brain must be unlike a normal egg. No single wax-like brain-core stores all types of information with a squishy periphery that achieves whatever else the mind can do. The brain must have two distinct and independent yokes or memory cores, one for learning new facts and events, and another for acquiring new muscle movement procedures, skills, and habits. Henry had split Humpty Dumpty in two.

Milner hypothesized those two memory systems but did not name them. In subsequent years they acquired labels such as the *declarative* versus *procedural* systems, the *what* versus *how* systems, and the *explicit* versus *implicit* systems.

However, Milner's research with Henry motivated the hypothetical properties of the two systems. One system was fast and conscious. It normally engages mechanisms in the hippocampal region without the need for repetition to store new facts and events. This seemed to explain the relation between Henry's hippocampal damage and his inability to learn new facts and events. For example, Henry never learned the *fact* that his early-onset osteoporosis (a bone disease) was caused by an anti-seizure drug that he had taken every day for decades (a series of *events*).[3]

Milner's habit memory system seemed different. It was slow and unconscious. It gradually built up *motor* procedures or muscle-movement memories via repetition, presumably in some yet-to-be-identified brain region untouched by Henry's operation. Those hypothetical processes are now called the *procedural learning hypothesis*.

THE PROCEDURAL LEARNING HYPOTHESIS

The procedural learning hypothesis seemed to explain how Henry succeeded in acquiring a variety of new muscle-movement memories after his operation. For example, beginning in his fifties, Henry needed to get around with a walking frame because his osteoporosis made him unsteady on his feet.[4] He had to learn to lean forward and push on the walker instead of balancing on one foot then the other as in normal walking. Henry easily acquired those new muscle-movements—presumably without awareness and via repeated trials, using his intact *procedural* system.

Milner's procedural learning hypothesis migrated well beyond mirror-tracing. The idea of muscle-movement memory eventually became a mantra. Sports psychologists advised professional basketball players "to let your muscles shoot the free-throws, not your mind."

Later experiments revealed other genuinely unconscious memory phenomena that seemed broadly analogous to procedural learning. The *word frequency effect* discussed earlier was one. Consistent with the procedural memory idea, frequent use gradually makes words easier to perceive, comprehend, and produce, an unconscious learning phenomenon discovered in the 1890s and replicated in hundreds of experiments since then.

Implicit learning was another phenomenon that seemed to reflect procedural learning. A classic experiment showed that Henry could learn implicitly—without awareness of learning—in a paradigm known as *repetition priming*. Henry and normal participants saw a series of words and responded *yes* for words containing the letter *A*, but *no* otherwise. They thought their ability to detect *A*s was being tested.

Actually, the real test came later in a supposedly unrelated experiment where the participants saw the first three letters of a word and had to produce as quickly as possible the first word beginning with those letters that came to mind. Shown the letters CON-, for example, the participants might report common words such as CONTROL, CONSIDER, CONVINCE, or CONNECT. An uncommon word such as CONTRALTO virtually never popped into awareness first. However, if participants, including Henry, saw CONTRALTO in the earlier *A* detection task, they often completed CON- with CONTRALTO—without awareness of its occurrence in the prior experiment. When queried afterward, Henry didn't even remember the *A* detection task, let alone his recent encounter with CONTRALTO. This unconscious effect of prior exposure did not differ in magnitude for Henry versus the normal participants in that study. Henry's implicit learning was intact. By Milner's logic, some intact brain system situated outside Henry's hippocampal region, say, the basal ganglia,[5] must have mediated this unconscious effect of prior exposure on Henry's memory for CONTRALTO.[6]

There were sceptics. Some researchers doubted whether mirror-tracing was a truly representative skill. Before accepting the procedural learning hypothesis, shouldn't Henry be tested on more typical skills, say, learning to read, write and speak a foreign language?

Other researchers questioned whether Henry's implicit learning reflected memory for *muscle movements*. After all, a phenomenon known as *semantic priming* is clearly *unrelated* to muscle movements but emerges from the same implicit learning paradigm as *repetition priming*. Participants in *semantic priming* experiments also perform an *A* detection task followed by the supposedly unrelated task of completing word-stems as quickly as possible with the first word that comes to mind. Shown the word-stem DOC-, they might say DOCKET or DOCKYARD. However, if a group of new participants responds to the word NURSE in the *A* detection task, and another group responds to the word CURSE, the NURSE participants will complete DOC- with the conceptually similar DOCTOR much more often than will the CURSE participants. Meanings rather than muscle memories clearly underlie this highly reliable *semantic priming* effect.

PUTTING HUMPTY DUMPTY TOGETHER AGAIN: CHALLENGES

From a scientific point of view, the procedural learning hypothesis suffered serious problems that took decades to discover. It was alleged to explain how normal individuals learn skills such as riding a bicycle, playing tennis, and skiing down a mountain.[7] It did not. It didn't even fully explain how people (including Henry) accomplished the artificial task of mirror-tracing around a star.

One unsolved problem concerned Henry's mirror-tracing times and errors. In the 1960s, Dr. Milner reported that Henry learned to mirror-trace stars extremely slowly. He required almost ten times as many practice trials to achieve the same level of proficiency as memory-normal controls.[8]

It wasn't that normal individuals became faster by straying into the error zone more often. Control participants both traced faster *and* made significantly fewer tracing errors than Henry. Nor could across-the-board slowing explain Henry's slower tracing times. Subsequent research showed that Henry was slower than normal controls only for unfamiliar tasks and stimuli, not familiar ones.[9] The procedural learning hypothesis focused on the *fact* that Henry learned to mirror-trace. It never adequately explained the nature of Henry's *slowness* in doing so.

Results of brain imaging studies further challenged the procedural learning hypothesis. In one recent study, researchers at Concordia University in Montreal imaged the brain activity of normal participants as they learned a novel motor skill that somewhat resembled typing or playing the piano.[10] The procedural learning hypothesis predicted heightened activity in a single localized *motor movement* area such as the basal ganglia. That did not happen. When participants acquired this and other novel skills, many widely separated areas in the brain became highly active, including the hippocampus and cortical areas with no known links to *motor movement*.

Adding to the challenge, hippocampal activity early in practice predicted the eventual degree of skill learning in the Concordia study. Participants with greater hippocampal activity on the first few trials of practice improved more by the end of the experiment than did participants with less hippocampal activity early on. Henry lacked that hippocampal contribution to improvement with practice in mirror-tracing. Is that why he learned to mirror-trace so slowly? The procedural learning hypothesis had no answer.

These brain imaging results raised other questions. Is mirror-tracing really a matter of learning *muscle* movements? What underlying research strategy motivated the procedural learning hypothesis in the first place? The remainder of this chapter describes that fundamental strategy: *the brain-first approach*.

THE BRAIN-FIRST APPROACH TO BRAIN AND BEHAVIOR

A standard strategy that I call the brain-first approach shaped and gave rise to the procedural memory hypothesis. The approach has four steps summarized below:

Brain-first Step One: Specify the impaired versus spared brain regions in a patient;
Brain-first Step Two: Determine what abilities of the patient are impaired versus spared;
Brain-first Step Three: Assign impaired abilities to damaged brain areas and spared abilities to intact brain areas;
Brain-first Step Four: Develop a theory of how neurons in different brain regions contribute to normal behavior.

THE BRAIN-FIRST APPROACH ILLUSTRATED

Case H.M. nicely illustrates the importance of Brain-first Step One. From 1953 until 2015, researchers spared no expense in trying to determine precisely which areas of Henry's brain were impaired versus spared. Before and after Henry's death, progressively more sophisticated imaging studies checked and rechecked the extent of Henry's 1953 brain damage. The outcome: previous conclusions more or less held up. The X-ray technology that William Scoville used to describe Henry's brain lesion proved reasonably accurate.[11]

Henry's story also illustrates Brain-first Step Two. Milner gave Henry a variety of cognitive tests to determine which of his abilities were normal versus impaired. This step appeared to bear fruit when Milner reported that Henry's memory tested normal in *motor* or *eye-hand coordination* tasks such as mirror-tracing around a star. Other aspects of Henry's memory that seemed intact included long-term memory for vocabulary and recall of events from short-term memory (up to thirty seconds later). We now know that these initial impressions were flawed.

Next came Brain-first Step Three—assigning Henry's impaired abilities to his brain damage and his spared abilities to intact areas of his brain. Based on research with Henry, researchers staked claims for separate brain areas that normally process short-term memories (intact), working memories (intact), long-term memories for facts and events (damaged), and procedural or motor-skill memories (intact). To date all of these claims remain controversial, perhaps because brain-first Step Three is still in its infancy. This leaves the still unborn Step Four. No researcher has so far explained findings with Henry in a detailed theory of how procedural memory, working memory, short-term memory, and long-term memory for facts events operate in the normal brain. The next chapter discusses a contrasting research strategy with results that support a surprising conclusion: the brain-first approach was wrong about Plato. In an important sense, Plato was right.

QUESTIONS FOR REFLECTION: CHAPTER 9

1) Why do you think that we are born to be forgetters as well as rememberers? Can you think of any advantages to not remembering everything?**
2) In the five-pointed star mirror-tracing task, why is the mirror important?

TEST YOUR MEMORY FOR CHAPTER 9

1) How accurate or inaccurate is the metaphor of *muscle memory* given what you now know about how your brain and memory work?**
2) What aspect of Henry's performance on the star mirror-tracing task inspired psychologists to split memory into two different systems (declarative vs procedural, implicit vs explicit)?
3) Why was Henry's extreme slowness early on in the mirror-tracing task problematic for the procedural learning hypothesis?

MEMORY TEST ANSWERS: CHAPTER 9

1) p. 134
2) p. 133
3) p. 136

Chapter 10

PUTTING HUMPTY DUMPTY TOGETHER AGAIN

DOES REPETITION *REALLY* MAKE THE HEART GROW FONDER?

The previous chapter discussed how the brain-first approach to studying Henry's mind and brain divided Plato's memory-wax into systems variously labeled *declarative* versus *procedural*, *what* versus *how*, and *explicit* versus *implicit*. Evidence developed under the fundamentally different *feat-first approach* that my UCLA lab adopted demonstrated a need to reassemble Humpty Dumpty into a unified theory of mind and brain.

Like the brain-first approach, the feat-first approach works in four standard steps:

Feat-first Step One: Describe in detail some feat-of-interest that the normal human brain achieves.

Feat-first Step Two: Develop a tentative theoretical account of how the normal brain performs the feat.

Feat-first Step Three: Test and refine that theoretical account by having normal participants perform the feat in well understood laboratory situations.

Feat-first Step Four: Develop a theory of how the normal brain performs the feat and how it breaks down with specific types of brain damage. Note that this step corresponds to the final, as yet unfulfilled step in the brain-first approach.

Using the feat-first approach to studying Henry's mind and brain, our initial focus was not on Henry's damaged brain but on the easily observable behavior of the normal people to be compared with Henry. Our starting question was not, *Why did Henry's brain mangle sentences and misremember facts and events?* but rather, *How does the normal human brain accomplish feats-of-interest such as producing coherent sentences, comprehending visual scenes, and acquiring procedural skills, including everyday skills such as riding a bicycle?* Our answer to that question led naturally to questions such as these: *What neurons in the brain accomplished Henry's intact feats-of-interest, and how? Did Henry's brain damage cause particular feats-of-interest to break down? If so, how exactly? If not, why not?* By asking those questions, my lab eventually developed a unified theory of mind and brain that clarified what *procedures* are and how they work.

THE FEAT-FIRST APPROACH IN ACTION

Explaining how the normal human brain learns to ride a bicycle was one of the ultimate goals of the procedural learning hypothesis. So let's discuss the feat-first approach to that problem before addressing the circus trick of mirror-tracing.

To understand bicycle riding feat-first, one must describe in detail the procedures underlying that everyday skill. Let's start with the simplest case: the safe procedures for learning to ride a gearless bicycle with hand brakes (see Illustration Box 10.1).[1]

Illustration Box 10.1: Basic Procedures for Learning to Ride a Bicycle

1. **Prepare Your Equipment**. Lower your bike seat until your feet can touch the ground when seated. Wear shoes that can stop the bike painlessly if you panic and try to brake with your feet. Ensure that your clothes and accessories cannot catch in your chain or spokes.

2. **Find a Suitable Practice Area.** Your practice zone should include a long flat space plus a gentle slope (to support momentum). Avoid steep hills, tight squeezes, or traffic of any kind. If concerned about falling, avoid concrete or pavement. Find a field with close-cropped grass.
3. **Mounting, Balancing and Steering.** Mount your bicycle, planting both feet on the ground. While seated, push yourself with your feet along your flat surface. Practice steering with only your handlebar. Notice how your body maintains balance by leaning slightly into a turn. When ready, practice making turns with your body rather than your hands. Notice how your front wheel automatically turns toward whatever direction you shift your body weight. Repeat until you can confidently steer the bike.

   Practice pushing yourself faster, then "gliding" with both feet in the air. Notice how balancing becomes difficult when moving too slowly. Move faster to help maintain balance.
4. **Practice Pedaling and Braking.** Practice starting and stopping on your long flat area. Begin from a complete stop with one foot on the ground and the other on its pedal at about ten o'clock. Push down on that pedal while lifting your other foot onto its pedal. Now push each pedal in alternation. You'll move. About thirty feet from the end of your flat area, brake gradually. Then dismount slowly.
5. **Practice Gliding down Slopes.** Mount your bike with one foot on the ground at the top of a gentle slope that ends in your flat area. Glide down the slope, gradually slowing to a stop in the flat zone at the bottom. Dismount, walk your bike back up the slope, and repeat several times to train your sense of balance. Repeat several times more while steering slightly left or right, then back to straight.
6. **Practice Gliding down the Slope and Pedaling on the Flat.** Glide down the slope and start pedaling along the flat surface. Make gradual turns, then sharper ones. Brake to a complete stop, balancing with only one foot.
7. **Practice Pedaling Up- and Down-hill.** Practice biking from your flat area up and down the slope several times. Note that moving uphill demands much more work. For

extra power, lean forward and stand on your pedals. To train for the more challenging terrains you will encounter in the future, practice pedaling halfway up the slope, stopping without dismounting, then continuing to the top.

8. **Practice Power-Riding.** Raise your seat to where you can touch the ground with one toe when seated. Note how this new configuration enables extra power. Practice power-riding on the flat until confident.

Note that procedures in Illustration Box 10.1 do not specify muscles or muscle movements. For adults, cognitive levels above the muscles represent most of what's new in learning to bike. Children form internal representations of the muscle-movements for actions such as *push down* when learning to climb stairs—long before learning to bike.

Note also that procedures such as *push down* only vaguely hint at actual movements. In pedaling a bicycle, the foot *simultaneously* pushes down as in climbing stairs *and* rotates at the ankle as in running. After successfully combining those familiar movements for pushing and rotating, bikers can forget the original hints for learning that aspect of riding a bicycle.

Needless to add, not everyone learns to cycle using the eight safe procedures outlined in Illustration Box 10.1. When I learned to bike at age five in northern Canada, my instructor skipped the *proper footwear* and *flat practice area* steps. Accelerating downhill, I tried without success to brake with my bare feet, then panicked. I swerved at high speed and could have suffered serious injury in the resulting tumble. Fortunately, I survived unscathed except for a life-long scar where a bare metal handlebar penetrated my right cheek.

But let's assume you avoided mishaps like mine and learned to bike without skipping any of the procedures in Illustration Box 10.1. What can you remember of those procedures now? People forget unused memories. If you adjusted your bike seat long ago, you probably forget the procedure for determining proper seat-height. Maybe you only remember the massively reiterated final procedure—power-biking. Frequently repeated procedures become unconscious but not forgotten.

However, at *some* point during learning, even your forgotten procedures must have entered awareness. Had you been unaware throughout, your unsafe shoes and loose clothes or trappings would probably have caught in your chain or gears or spokes and caused a memorable accident somewhere along the way. The procedural learning assumption that bike-riding procedures *never* enter consciousness no longer seems plausible.

Of course, *artificial* skills might be different. Learning the *unnatural* skill of mirror-tracing without instruction might be entirely unconscious. Perhaps the procedural learning hypothesis is right that procedures *never* enter awareness when people trace a figure seen in a mirror that reverses up and down.

To find out, I conducted feat-first analyses of the procedures for mirror-tracing a star and discovered that stars are less than ideal for testing awareness during procedural learning. The many oblique angles in a star are difficult to trace and describe. If participants become fleetingly aware of the procedures for mirror-tracing a star they might later be unable to describe that awareness to an experimenter. I decided to include an additional figure in our awareness experiments: squares. A figure with simpler or easier-to-describe mirror-tracing procedures than a square is difficult to imagine (see Illustration Box 10.2).

Illustration Box 10.2: PROCEDURES FOR MIRROR-TRACING A SQUARE

To move the pencil in the mirror . . .

1. . . . vertically down, move your hand up;
2. . . . vertically up, move your hand down;
3. . . . horizontally left, move your hand left;
4. . . . horizontally right, move your hand right.

Before entering the laboratory, people unconsciously know square-tracing procedures three and four in Illustration Box 10.2. In normal everyday life, we move our hand leftward to trace left and rightward to trace right. Exactly the same rules apply when mirror-tracing the horizontal arms of a star. If the procedural learning hypothesis is correct about learning to mirror-trace being entirely unconscious, participants should be unaware of the overlearned procedures for horizontal hand movement when mirror-tracing, just as when moving their hands horizontally in daily life.

To test this procedural learning prediction, my lab first developed a tentative theoretical account of how the normal brain mirror-traces, then refined that account in pilot tests with well understood laboratory tasks. I'll illustrate the important qualifier *well-understood* shortly. For now, however, note that applying the labels *motor* or *muscle movement* to the mirror-tracing task does not make it *well understood.* Those modifiers are themselves hypotheses. To determine whether learning to mirror-trace engages only low-level *motor* processes for moving the muscles requires an experiment.

THE UCLA MIRROR-TRACING EXPERIMENTS

Results of three experiments in my UCLA lab showed that learning is neither entirely unconscious nor restricted to acquiring muscle movements when normal people mirror-trace stars and squares. Our first study replicated Milner's star-tracing experiment discussed in the previous chapter, but we added four significant changes: we tested for awareness of the star-tracing procedures immediately after the experiment; our participants were sixty UCLA undergraduates without hippocampal damage; a computer replaced Milner, her paper, her mirror, and her stopwatch; and the dot-cursor on the computer monitor replaced Henry's pencil marks. Like Milner, the computer recorded and saved participants' response times and errors. Like the mirror in Milner's experiment with Henry, our computer also reversed up and down on its monitor, displaying the trajectory of participants' hand-mouse movements in real time as a solid line superimposed over the star pattern.

Without mentioning mirrors, instructions warned participants that some of the normal rules relating mouse movement to cursor behavior had been reprogrammed. The participants then used the mouse to move the cursor counterclockwise around the star as quickly as possible back to the starting point.

Half the participants traced the star pattern thirty times and half traced it only once. All participants then answered two-choice *awareness questions* about how they performed the task. Examples are: *How did you move your hand to shift the cursor horizontally to the right?* (correct answer: *rightward*), and *How did you move your hand to shift the cursor vertically upward?* (correct answer: *downward*).

The results showed that virtually every participant was aware of the four mirror-tracing procedures described in Illustration Box 10.2. Regardless of the extent of prior practice (one versus thirty practice trials), participants correctly answered the awareness questions for vertical and horizontal movements significantly more often than if they had been responding randomly (based on tossing a coin). Contrary to the procedural learning hypothesis, procedures are *conscious* when people learn artificial as well as common skills.

A follow-up experiment replaced the star with a square and replicated these basic awareness results. "Wait," emailed a colleague after reading the preceding sentence. "Why the identical results? Aren't stars more complex than squares?"

"Those matching findings make sense," I replied. "The rules underlying star- and square-tracing are identical, but applying them to stars is complex. For each line in a square, only one of the mirror-tracing rules in Illustration Box 10.2 applies. Stars are different. To mirror-trace down the left flank of the leftward pointing leg of a star, for example, one must apply *two* of the rules in Illustration 10.2: move your hand up somewhat to shift the cursor vertically down *and* move your hand left somewhat to shift the cursor horizontally left. To successfully navigate eight of the ten lines in a five-point star, one must *simultaneously and to different extents* combine two of the mirror-tracing rules in Illustration Box 10.2. This complexity explains why Milner's participants required one to three days of practice to mirror-trace a star without errors. What a contrast with our UCLA participants who required only minutes to mirror-trace a square without errors."

Our square-tracing study prompted other new insights. Awareness of the proper procedures early in practice correlated with better performance late in practice. Participants who correctly answered our awareness questions after the first trial of square-tracing responded faster and with fewer errors on the thirtieth trial. Consistent with results of the Concordia study discussed in the previous chapter, awareness of the appropriate procedures early in our skill-learning task facilitated performance throughout the task. Contrary to the procedural learning hypothesis, awareness not only happens during skill learning, it helps.

There was another interesting twist. Conscious *recall* of the correct procedures *diminished* over time. Participants who indicated awareness of the appropriate tracing procedures early in the experiment often responded, "I forget," when tested again at the end of the experiment. They could no longer recognize the procedural *descriptions* underlying their successful performance of the task! Just as forgetting wipes out the once conscious but no longer needed verbal descriptors when learning to ride a bicycle (see Illustration Box 10.1), so too with learning to mirror-trace a square.

My lab next tackled the procedural learning assumption that mirror-tracing is a strictly *motor* or *muscle-movement* skill. We asked a group of experimental participants to mirror-trace a star with their *non-preferred* hand for thirty trials, then to continue mirror-tracing for ten trials with their *preferred* hand. If skill learning takes place entirely within the muscles as per the procedural learning hypothesis, then mirror-tracing improvements with the first hand should not benefit subsequent performance with the completely different muscles of the other hand.

That is not what happened. The experimental group performed as well as a control group that mirror-traced the star forty times with their preferred hand. Response times and errors on the fortieth trial were virtually identical for the experimental and control groups. This means that learning in both groups occurred not in the *muscles*, but at a *higher, cognitive level in the brain*. The data indicate that mirror-tracing is a *mental* skill like chess. Just as the hand muscles for moving a chess piece cannot force checkmate to win the game, so too with square-tracing. A brain area well above the muscles learns the concept *move your hand up to shift your pencil down.*

HUMPTY DUMPTY COMES TOGETHER AGAIN

Brain science now knows the what and where of Plato's memory-wax. It is distributed throughout the brain rather than confined to a particular region such as the hippocampus. It resides in the biochemistry that occupies the trillions of tiny "gaps" or synapses that separate the billions of connected neurons in the brain.

Any synapse, anywhere in the central nervous system of the brain, has a *pre-synaptic neuron* that initiates the process of activating its *post-synaptic neuron* on the far side of the gap. When this post-synaptic neuron becomes activated, neurochemicals in their shared synapse undergo a slight but lasting change that facilitates activation of the post-synaptic neuron via that synapse in the future. These microscopic chemical changes inside the synapse are the basis for every type of memory, including internal representations for facts, events, and procedures.

To illustrate how synaptic changes influence behavior, consider the fact that frequent use makes words easier to perceive, comprehend, and produce. Whenever speakers use a word, they activate a hierarchy of connected cortical neurons that represent its meaning, its syllables, and its speech sounds. With repeated activation of the hierarchy, the synapses linking its neurons become more conducive to future activation. These miniscule synaptic changes induced with each activation gradually accumulate over the lifetime of the speaker, making frequently used words easier to perceive, comprehend, and retrieve, all without conscious awareness. The same use-it-and-improve-it principle applies to every memory that can be repeatedly retrieved—words, facts, events, concepts, and images.

WHAT ABOUT YOUR HIPPOCAMPUS, HENRY, AND OURS?

If chemical changes in synapses resulting from repeated activation of post-synaptic neurons are the basis for all learning and memory, what does the hippocampus do? Why was Henry able to create memories for novel mirror-tracing actions but not for novel facts and events? What follows is an *integrated learning hypothesis* that answers these and many other questions.

The hippocampus contains binding mechanisms that powerfully activate two or more neurons in the cortex that represent novel or never previously activated conjunctions of information. In the previous sentence, the noun phrase *binding mechanisms* represents a novel conjunction of information for readers who have never previously used or encountered the word *binding* combined with *mechanisms*.

Like browsers, *hippocampal binding mechanisms* are powerful drivers that ramp up the activation level of connected cortical neurons. However, browsers are weak drivers that ramp up the activity level of connected neurons only moderately higher and for brief periods (measured in hundredths of a second). Hippocampal binding mechanisms drive the activation rate of connected neurons to extremely high levels (up to 1500 activation cycles per second) over extended periods (measured in seconds).[2] By inducing this prolonged, intense activity, the hippocampus alters the biochemical memory-wax in the synaptic junctions of those cortical neurons at a critical time—early in learning when new representations are being formed in the brain.

The news then is not that Plato got it wrong about experiences leaving image-like impressions on a waxy substance in the brain that can later be used to reconstruct the original experience. Even modern scientific metaphors get the details wrong in specifiable ways.[3] The surprise is that Plato was right in an important way. A pliant substance in the brain *does* represent memories! As with Plato's memory-wax, a single process—repeatedly triggered activation of post-synaptic neurons—alters the chemistry of synaptic memory-wax everywhere in the brain, regardless of the type of information a synapse encodes. The hippocampus simply accelerates the process. In the integrated learning hypothesis, repeated neural activation underlies all learning, whether fast or slow.

Problems that were unsolved in the procedural learning hypothesis disappear in the integrated learning hypothesis. When people acquire a new or unfamiliar skill, the hippocampus speeds up the learning process. This explains why the hippocampus becomes engaged early in practice when learning a novel skill and why enhanced hippocampal activity early in practice predicts improved performance late in practice.[4] This also explains why Henry needed more practice trials than normal to become proficient at mirror-tracing. For normal individuals, the hippocampus accelerated the rate of learning to mirror-trace. Lacking a hippocampus, Henry could only learn gradually, via repeated activation, one practice trial at a time.

The integrated learning hypothesis also promises insights into the mystery of consciousness. A later chapter describes how it explains a variety of observed relations between awareness and learning.

The fact that practice facilitates both cognitive and muscle-movement memories also makes sense under the integrated learning hypothesis. Repeated activation of post-synaptic neurons strengthens internal representations for any type of information, muscle-movement skills as well as cognitive skills such as perceiving the visual world.[5] The next chapter examines those visual perception skills through the lens of H.M., the man who mistook a wastebasket for a window.

WHY DOES REPETITION MAKE THE HEART GROW FONDER?

The integrated learning hypothesis summarizes, explains, and predicts relations between the brain, repeated activation and skills such as learning, comprehending, and producing words. It's just a starter kit. If the hypothesis survives extensive tests, extending it to other kinds of repetition phenomena might be possible. One example is the way repetition makes the heart grow fonder. Novel stimuli initially elicit fear and avoidance, especially in children and animals, but repetition transforms these negative reactions to positive. Through repeated encounters, strangers seem to become more likeable, pleasant, and attractive. So do unfamiliar paintings, Chinese characters, and nonsense words.[6]

Repetition also breeds desire and belief. Business executives know that massively repeated advertisements, although annoying, can increase the probability of a sale when people later see the advertised object in a store. Political leaders in Nazi Germany also discovered a link between repetition and belief. People often come to believe false, outrageous, or ridiculous messages that an authoritative source such as the government frequently reiterates. Officials in the George W. Bush administration may have known this when they repeatedly linked Saddam Hussein with September 11, 2001. More than 40 percent of Americans falsely concluded that Iraqi citizens were involved in the 9-11 attacks when, in fact, none were.[7] Finally, repetition can breed distortion and meaninglessness, as in the verbal transformation and semantic satiation effects. If the word *police* is replayed for an extended period via audio tape-loop, people hear illusory forms such as *please*, *fleas*, *trees*, and *fleece* that eventually seem to lose their meaning. Try it. Quickly repeat the word *police* silently to yourself for a few minutes and see whether its original meaning seems to change.[8]

QUESTIONS FOR REFLECTION: CHAPTER 10

1) When trying to understand how your brain generates some behavior, how does the feat-first approach help overcome some problems of the brain-first approach?
2) We see that "repetition makes the heart grow fonder." But have you also experienced the opposite effect, where repetition makes the heart grow colder? How could the same process, repetition, produce opposite effects? Can you use the integrated learning hypothesis to suggest an answer?
3) How have you taken advantage of the use-it-and-improve-it principle in your own life? What would inspire you to employ that procedure more often?**
4) When learning how to perform complex actions, if you are aware of the correct procedures early in the learning process, then your subsequent performance improves even if you cannot subsequently recall the procedures. Would your performance improve even more if you reminded yourself of the procedures in order to prevent forgetting?

TEST YOUR MEMORY FOR CHAPTER 10

1) What is the use-it-and-improve-it principle and how is it relevant to your memory for all kinds of words, events, concepts, and images?
2) How does your hippocampus and its binding mechanism allow you to create novel combinations of words and remember those word combinations again?
3) How would you explain the integrated learning hypothesis to somebody who is worried about and trying to prevent memory loss?**
4) What evidence indicated that early awareness of proper procedures predicts subsequent success in the mirror-tracing task?
5) What evidence indicates that what trainers call *muscle memory* may not actually be based on muscles at all?
6) Plato thought that memory-wax was the basis for all memory. Was he right?
7) How does the hippocampus facilitate procedural learning?

MEMORY TEST ANSWERS: CHAPTER 10

1) p. 149
2) pp. 150–51
3) pp. 150–51
4) p. 148
5) p. 148
6) p. 149
7) pp. 151–52

SECTION III
AWARENESS OUT OF THE BLUE

New observations, insights, and ideas often come to us out of the blue, a mysterious region of the brain that responds to the unexpected and welcomes novelty into the mind with a whispered *aha.* Section III focuses on this mysterious brain region.[1]

Chapter 11
CAN YOU CREATE NEW CONCEPTS, HENRY?

WHAT WOULD HAPPEN IF YOU LOST YOUR INBORN ABILITY TO CREATE?

Scientific discoveries often arrive as unexpected visitors—out of the blue. That was how I discovered the connection between creativity, ambiguity, and the hippocampus. My first surprise visitor emerged from the blue in summer of 1966. I was a young graduate student working in a vacant office in William James Hall at Harvard University. Twenty Harvard undergraduates came to visit me one at a time. They faced me across my desk as I explained the task: to detect and describe the two meanings in ambiguous sentences as quickly as possible. To illustrate, I described the two meanings of a typical sentence. "*We are confident that you can make it* can mean either *to construct a physical object* or *to attend some event*," I said. Beside me was a stack of thirty-two sentences typed on four by seven inch index cards.

To begin the experiment, I started a timer and simultaneously flipped over the top card to expose the first ambiguous sentence. I stopped the timer if participants said *yes*, indicating detection of a second meaning, *or* if they *failed* to detect a second meaning in the sentence within ninety seconds. For the latter participants, I explained both meanings of the sentence before proceeding to the next trial.

As it turned out, I only needed to enforce the ninety-second cut-off rule with Henry. The undergraduates invariably said *yes* within a few seconds and briefly described both meanings of the sentence in the order that they saw them. In a later experiment, for example, *To get somewhere or to make something* was how the undergraduates typically described the ambiguity in *We are confident that you can make it.*

After each trial I recorded which meaning the participant reported first, plus their time to say *yes.* I believed those detection times would help solve a deep mystery that was just beginning to surface when I was a graduate student. Susimo Kuno—a professor of Linguistics at Harvard—had recently demonstrated that natural language was much more ambiguous than anyone realized back then. Dr. Kuno had developed a computer program that included an elaborate *transformational* grammar (a new concept at the time) plus a digital lexicon that was more extensive than Webster's: it listed as many as forty-four different meanings for common words such as *like*, *make*, and *run*. The program was supposed to translate English sentences into concepts the way humans do. It did not. Asked to interpret the seemingly simple sentence *Time flies like an arrow*, Dr. Kuno's program spewed out eighty-three interpretations that people immediately recognize as absurd. One was: "Get out a stop watch and time flies in the same way an arrow times them." Another was: "There is a special kind of fly. Not house flies, sand flies, or horse flies, but *time* flies. And those time flies like an arrow." *Ambiguity* was clearly the Achilles heel of Dr. Kuno's program.

Why don't people generate equally ridiculous interpretations when comprehending everyday sentences? Answering that question seemed essential if computer programs were ever going to understand language the way humans do. Decades after the invention of digital computers, natural language comprehension by computers was only a dream when I was a graduate student.

That dream almost blinded me to an unexpected visitor: the word *aha.* The undergraduates in my ambiguity detection experiment often gasped a quiet *aha* right before they discovered the second meaning of a sentence. Questioned after the experiment, one student described her *aha* as reflecting a click of awareness, a sudden perceptual shift from fuzzy to lucid. Another compared his *aha* to Archimedes' *Eureka!*

The renowned German psychologist, Karl Dunker, had made the word *aha* famous in psychology during the 1930s. *Aha* was what his students at Swarthmore College said just before they solved his moderately difficult insight problems. His most famous problem went like this: "Here are some matches, a box of thumbtacks, and a candle. Your goal is to light the candle and attach it to the wall using only those objects."

To solve the problem, the students had to mentally transform the box of thumbtacks from a container into a basin that could be mounted on the wall to support the candle. That insight came as a sudden and gratifying psychological flip that triggered the *aha* reaction and the knowledge that the problem was solved. All that remained was to translate the idea into actions: remove the tacks, tack the empty box to the wall, light the candle, and drip wax onto the floor of the box to hold the candle upright.[1]

Those *aha*s suggested that detecting linguistic ambiguity and solving Dunker's insight problems engage similar processes. However, my Eureka moment came when another visitor from out-of-the-blue revealed for me the fundamental brain mechanism underlying ambiguity detection.

That out-of-the-blue visitor was Henry. In the spring of 1966, Dr. Hans-Lukas Teuber, chair of the MIT Psychology Department, introduced me to Henry and suggested I test his ability to comprehend sentences using the ambiguity detection task I had given the Harvard undergraduates.[2] It never crossed my mind that Henry's performance on that task would alter the direction of my career and transform my understanding of memory, mind, and brain.

Henry was *very* unlike the Harvard students. Often I could not tell whether Henry saw one meaning or two in the ambiguous sentences. He kept using the ambiguous words to explain the ambiguities, despite my repeated requests not to do that. Likewise against my instructions, Henry immediately gushed out whatever meaning came to mind. He failed to say *yes* to mark when he discovered a second meaning. Not that it mattered. Henry virtually never succeeded in comprehending both meanings of the sentences on his own.[3] Usually he simply used different words to redescribe the same meaning—as in this excerpt:

Henry [explaining the two meanings of *We are confident that you* *can make it*]: "Well, that is like, um, people are sure that a person can do something. And also I thought of, that a person is sure himself that others are sure that he can do it."

Henry reiterated that same interpretation seven times without ever describing a second interpretation for that sentence. Even when asked to stop doing that, Henry immediately repeated himself, as if oblivious to the request. Henry couldn't even repeat the sentence interpretations that I explained to him after the ninety second time limit.

Also abnormal, Henry often interpreted the sentences in ungrammatical or impossible ways. In the prior excerpt, "A person is sure himself that others are sure that he can do it" is an impossible interpretation of *We are confident that you can make it*. The *we* in *We are confident* cannot mean *a person himself*. The *you* in *you can make it* cannot mean *he*.

Grammatical errors also abounded in Henry's utterances. When describing a single sentence, Henry used twenty pronouns in ungrammatical ways. By contrast, none of the Harvard undergraduates misused even one pronoun throughout the entire experiment.

Equally abnormal was the way Henry peppered his utterances with hundreds of filler words and clichés: *naturally, well, in a way, I thought of, You'd call it, I'm having an argument with myself*. These reiterated bromides spoke to the demise of Henry's creativity.[4] So did the missing clicks of comprehension. When I explained a sentence meaning that Henry failed to discover on his own, Henry typically responded not *Aha! Sure enough!* but, "I wonder," as if he did not really understand.

Henry's behavior surprised me as a twenty-three-year-old graduate student. I could see that his sentence comprehension and production were impaired. I just could not imagine why. I only came to understand the relation between language, problem solving, and hippocampal damage many years later.

I nonetheless considered my 1966 observations important enough to refine. In a follow-up study at UCLA, I gave my ambiguity detection task to a group of memory-normal individuals who precisely matched Henry in age, IQ, education (a high school degree), and background (semi-skilled labor). That group performed as well as the Harvard undergraduates, ruling out age, IQ, education, and background as possible reasons for Henry's deficits in my 1966 ambiguity detection study. Similar matching procedures, although expensive and time consuming, became part of every subsequent study that I conducted with Henry.[5]

As soon as I could test Henry himself again, I also ruled out forgetting as the basis for his problems in detecting multiple meanings. In collaboration with my UCLA post-doctoral fellow, Dr. Lori James, I created an ambiguity detection task where nothing had to be remembered. On each trial in this task, Henry and the memory-normal controls received a card displaying a target sentence plus a single interpretation that was correct for half the sentences (for example, *The umpire unexpectedly called the pitch a strike* for *When a strike was called, it surprised everyone*) and incorrect for the other half (for example, *The umpire quickly called the coaches to the mound* for *When a strike was called, it surprised everyone*). As in my 1966 study, the target sentences were always ambiguous, but this time, the instructions did not mention ambiguity. All Henry had to do was say *yes* if the interpretation fit the sentence and *no* otherwise.

A day later, Dr. James reran the study with Henry but coupled each ambiguous sentence with its *alternate* correct interpretation. If the correct choice for *When a strike was called, it surprised everyone* on day one was the interpretation, *The umpire unexpectedly called the pitch a strike.* Then on day two, *The union workers unexpectedly went on a labor strike* was the correct interpretation. The day two results showed that Henry performed close to chance (50 percent) and no better than on day one. Unlike my matched memory-normal individuals, Henry could not tell the difference between correct versus incorrect interpretations of an ambiguous sentence when he didn't have to remember one meaning while searching for the second. Forgetting could not explain my 1966 results.[6] I could forget about forgetting.

Again with the help of Dr. James, a final experiment demonstrated what kind of problem solving ambiguity poses. This study asked Henry to detect the two meanings of ambiguous *words* presented either *in isolation* or *in sentences*. The isolated ambiguous words came from the ambiguous sentences in my 1966 experiment with Henry. For example, we simply extracted the ambiguous word *tank* from the sentence *The soldier put the gasoline in the tank*, handed it to Henry on an index card, and asked him to detect and describe its two distinct meanings.

Henry easily detected and described the meanings of isolated ambiguous words. Gone were his difficulties in detecting and describing the same ambiguous words in sentences.[7] To get a feel for this remarkable difference, compare how Henry described what *the position* means in isolation (a) versus in the sentence *The marine captain liked the new position* (b).

(a) **Henry** (describing two meanings of *the position* in isolation): The position ... could be the place where you work, have work. And also it could be your election place. Spot. Position. (Presumably Henry's "election ... place ... spot" refers to the *spot* or *place* where you elect to be)

(b) **Henry** (describing two meanings of *the position* in the sentence *The marine captain liked the new position*): The first thing I thought of was a marine captain he liked the new position on a boat that he was in charge of, the size and kind it was and that he was just made a marine captain and that's why he liked the position too, because he was above them and of all, most of all ...

Experimenter: So you're saying that he like [*sic*] his job in other words.

Henry: He liked his job.

Experimenter: Okay. Now, there is another meaning in that sentence. Can you tell me what it is?

Henry: I just gave you two.

Experimenter: Those are both really the same. Because they were both related to his job. There is another meaning.

Henry: Well, cause he was on a new boat, you might say a new boat, he was made captain of a new liner or whatever it is and it's different than what he had before. He might have had a ... a ... a ...

Experimenter: You mean his job was different.

Henry: Yes, he might, he has people ...

Experimenter: That's the same meaning that you told me. There's another meaning that's suggested by those same words, in that same order, something ...

Henry: Well, you see, I thought of marine captain in a new position, was one, was transporting goods, he was a marine captain of a boat there and then marine captain he liked the new position because of being, being a passenger line, I'd guess you'd call it, because of the people that would . . .

Experimenter: All right, I'm going to tell you what the other meaning is. One meaning, that one that you have, means that he likes his new job. The other meaning is that he likes his new physical position. In other words, he may have been standing up on watch for a great number of hours and then he gets to sit down and he likes that new position of being able to sit down. The position of his body.

Henry: Oh.

Experimenter: Okay? Do you see how those are really rather different meanings?

Henry: They're different.

Experimenter: One has to do with his job and the other is if he is sitting, standing, or whatever.

Henry: The position he's in.

Experimenter: The position of his body. OK, you see? Do you understand how the very same words can mean two different things, two different interpretations depending on how you read it? OK. [next trial]

As an isolated word, *position* means *a job, a location* or *to situate or arrange.* Henry learned that as a child. He could recall those preformed meaning representations via his intact retrieval mechanisms. Henry's intact retrieval mechanisms likewise allowed him to remember familiar facts such as "I was born in 1926."

Novel sentences are different, however. Already formed internal representations of what *position* means make sense in some sentence contexts but not others. *Position* can mean *to situate or arrange* in the context *The marine captain tried to position his troops on the hill* but not in the context *The marine captain liked the new position*. To accurately comprehend the two meanings of an ambiguous sentence, people must create new internal representations that coherently integrate the meanings of its words with their context of occurrence in the sentence. This Henry could not do after his operation because he could not create new internal representations. Henry's inability to remember new facts calls for the same explanation. After his operation, Henry could not remember his *current* age because that meant forming a new internal representation every year.

WAS YOUR CEREBELLUM GUILTY, HENRY?

In 2004, neuroscientists at the University of Utrecht in Holland discovered something that could have shattered my theory of ambiguity detection and the hippocampus. They showed that neural activity in the cerebellum, a structure below the visual cortex at the back of the brain, increases significantly when normal speakers comprehend the two meanings of ambiguous sentences. As in my ambiguity-detection experiment, participants in the Dutch experiment read ambiguous sentences silently and reported both meanings as soon as they discovered the second interpretation. The sole difference was the type of ambiguity: the Dutch researchers only presented sentences containing *structural ambiguities*, as in *Pavlov fed her dog biscuits*, where a woman is fed dog biscuits in one interpretation, and her dog is fed biscuits in the other.[8]

Their finding—heightened activity in the cerebellum—was relevant to the damage to Henry's cerebellum caused by the anti-seizure medicine he took for decades. I now needed to determine whether Henry's cerebellar damage or his hippocampal damage caused his deficits in comprehending ambiguities. My interpretation of his comprehension and memory problems depended on the answer.

To address this issue, I again proceeded feat-first: I compared what the intact cerebellum usually does with how people normally comprehend the two meanings of structural ambiguities—the type of ambiguity in the Dutch experiment. The routine job of the cerebellum is to control the timing of people's actions when they speak, walk, or move their eyes.[9] This link between timing and the cerebellum readily explains the Dutch results because timed pauses help people detect the two meanings of structural ambiguities. To see how, try this little exercise. Say the sentence *Pavlov fed her dog biscuits* silently to yourself with two different timing patterns: *Pavlov fed her* ___ *dog biscuits*, and *Pavlov fed* ___ *her dog*___ *biscuits.* Now try that again and notice how the sentence changes in meaning. First Pavlov feeds a woman some dog biscuits, then he feeds her dog some biscuits. Inserting *timed pauses* into your internal speech revealed the two meanings of that structural ambiguity!

To detect the dual meanings of structurally ambiguous sentences in the Dutch experiment, participants must have internally generated timing pauses—causing enhanced cerebellar activity. Consistent with this explanation, the Dutch team also observed heightened activity in cortical areas associated with internal speech as well as the cerebellum!

Likewise consistent with my timing explanation, research teams at UCLA and Oxford University discovered that amnesics *without cerebellar damage* experienced the same difficulties as Henry when detecting two meanings in *lexically* ambiguous sentences resembling *The soldier put the gasoline into the tank* (where *tank* can mean either a container or a military vehicle).[10] This finding is important because timing pauses only help in detecting *structural* ambiguities, not *lexical* ambiguities. To experience this difference, repeat *The soldier put the gasoline into the tank* four times to yourself, pausing after different words each time. Unlike before, your pauses will leave your interpretation of that lexically ambiguous sentence *unchanged*. This means that Henry's cerebellar damage cannot explain his deficits in comprehending *lexical* and perhaps other types of ambiguities.[11] However, an experiment discussed shortly provided the truly conclusive proof that Henry's hippocampal rather than cerebellar damage caused his comprehension deficits. My UCLA lab showed that patients *with* cerebellar damage but *without* amnesia comprehended ambiguous and unambiguous sentences without difficulty! With that demonstration, Henry's cerebellar damage no longer cast a shadow over my research on his language and memory.

WHAT ELSE CAN'T YOU COMPREHEND, HENRY?

Were Henry's comprehension deficits limited to ambiguous sentences? To find out, Dr. James and I tested whether Henry could distinguish grammatical from ungrammatical sentences. He could not. Handed a *grammatical* sentence such as *John gave me the car that he couldn't drive by himself*, Henry responded, "Yes, grammatical," significantly less often than a group of carefully matched control participants. Shown an *ungrammatical* sentence such as *John gave me the car that he couldn't drive by ourselves*, Henry responded, "No, ungrammatical," reliably less often than the normal controls.

For sentences with a single incorrect word, such as *ourselves* in *John gave me the car that he couldn't drive by ourselves*, Henry performed close to chance (50 percent) and significantly lower than the 88 percent average for normal individuals. Was Henry just guessing? To find out, my lab reran the study with a twist: after participants responded "No, ungrammatical," Dr. James asked them to specify what word was wrong or wrongly ordered, and then correct it. This was easy for the control participants. After identifying *ourselves* as incorrect, they quickly supplied a correction: *John gave me the car that he couldn't drive by himself.*

However, Henry called correct words incorrect and could not put right what he claimed was wrong. For *Will be Harry blamed for the accident*, Henry fingered *blamed* as the incorrect word, but insisted that correction was impossible without further information: "Well, you have to find out what Harry was blamed for," said Henry. "And it—possibly that word b—Harry is blamed, in a way, for the accident and he could be blamed for something else." Henry was indeed guessing when he called sentences ungrammatical.

Can inability to follow the instructions explain why Henry responded as though flipping a coin to determine his answers? The experiment contained four severely ungrammatical foils to rule out that possibility. The foils thoroughly shuffled the words in a grammatical sentence. *She has decided to buy a house* rearranged as *Decided has house she a buy to* did not fool Henry. He always recognized the foils as ungrammatical, indicating clear understanding of the frequently repeated instructions. Henry just couldn't comprehend "almost correct" sentences.

The memory-as-internal-representation perspective explains why Henry could not discriminate between grammatical versus ungrammatical sentences. To discover that *John gave me the car that he couldn't drive by ourselves* is ungrammatical, Henry had to create, and reject as incoherent, several new internal representations. If *he* refers to *John*, the sentence does not work unless it ends *that he couldn't drive by himself*; and if *ourselves* refers to *John* and *me*, the sentence is incoherent because it does not end *that we couldn't drive by ourselves.* Similarly, to recognize that *John gave me the car that he couldn't drive by himself* is grammatical, Henry had to create a new internal representation in which *John*, *he*, and *himself* refer to the same person. Due to his hippocampal damage, Henry could not form such new internal representations. However, he could recognize the ungrammaticality of random word strings where no new and coherent internal representation was possible.

QUESTIONS FOR REFLECTION: CHAPTER 11

1) If an especially challenging problem requires a sudden insight or *aha* moment for you to solve, do you feel like you would be more or less likely to remember the solution? Why might that be?
2) If you practice introducing ambiguous words into your speech, do you think that you will be more likely to detect ambiguity in other people's speech? Why or why not?
3) Why was it important that experiments compare Henry with control participants who are closely matched "on relevant factors other than brain damage"? Why does the scientific method in general depend on investigating effects of a single variable or factor while keeping constant all other relevant variables or factors?
4) Do you think that Henry would have problems understanding jokes? Why or why not?

TEST YOUR MEMORY FOR CHAPTER 11

1) If you somehow lost your ability to vary the pauses between words in your natural speech, how would this likely change the degree of ambiguity that others would detect in your speech when you talk?
2) If you lost your ability to form new internal representations like Henry did, how would this likely alter your speech, and your awareness of those changes?**
3) Why could Henry understand ambiguity in familiar words or phrases presented in isolation but not when those same words appeared in novel sentences?
4) Why was it important to ensure that cerebellar damage could not explain Henry's deficits in understanding lexical ambiguity?

MEMORY TEST ANSWERS: CHAPTER 11

1) pp. 164–65
2) pp. 166–67
3) p. 167
4) pp. 164–65

Chapter 12
CREATIVE COMPREHENSION IN EVERYDAY LIFE

HOW CAN YOU ENHANCE YOUR CREATIVITY AND MENTAL AGILITY?

As discussed in Chapter 11, research with Henry pointed to a connection between creativity, sentence comprehension, and the hippocampus. This chapter explores the relation between the hippocampus and creativity more broadly, beginning with the question: what is creativity?

Innovation and usefulness are the universally accepted hallmarks of creativity.[1] Every great work of genius is novel and valuable. Being new in the history of the world and helpful to humankind, the bifocal lens that Benjamin Franklin invented was undeniably creative. In the art world, Picasso's fresh and influential Cubist way of perceiving visual forms was also creative. So was Shakespeare's "Juliet is the sun." In the world of literature, that metaphor was new—nobody had previously characterized Juliet Capulet that way—and useful. It allowed Shakespeare's audiences to quickly grasp the pivotal idea in his play—that Romeo saw Juliet as warm, life-giving, and central to his universe.

However, the widely accepted idea that genius equals creativity equals novel-in-world-history is problematic. As a starter, that equation reflects an inaccurate and outdated stereotype about creativity, genius, and the mind. The stereotype originated in the eighteenth century. Thinkers in the Romantic tradition hypothesized that geniuses possess a fundamentally different kind of mind and brain from everyone else.

In the stereotype, geniuses generate powerful new ideas because their *unconscious* mind is especially quirky, productive, and *magical.* Geniuses were thought to consciously consider a critical problem and put it aside to let their unconscious deliver a fully formed solution out of the blue—perhaps in a dream. Interludes away from work were believed essential to help the unconscious mind of a genius incubate and solve a problem. God could also play a role. The creative insights of Leonardo da Vinci were considered divinely inspired "bolts from the unconscious." Great musical masterpieces were attributed to the voice of God speaking directly inside the unconscious mind of Mozart and Beethoven.

Historians have tried without success to verify that Romantic era conception of creativity. It turns out that geniuses were especially creative when introspecting about creativity in their autobiographies. Objective evidence from diaries, notebooks, draft manuscripts, and correspondence suggests that more mundane factors also contributed to their creativity. Genetics is one. Good fortune is another. Lucky prodigies stumble onto a fundamental problem that is new and solvable. Unlucky geniuses with equivalent talent and dedication are not remembered as geniuses. Hard work and a passion to excel also underwrite creative masterworks. In general, geniuses are industrious wonks. Before contributing creative products of lasting value, most geniuses immerse themselves in their field for at least ten years. During that period, they endlessly revise their thinking, gradually closing in on an ideal solution to their central problem. They typically work day and night, absorbing a vast array of strategies for solving tens of thousands of sub-problems. Their breakthrough insight usually tweaks an earlier attempt that failed. Mozart's first high-quality opus emerged after twelve years of hard work. The symphonies he composed at age eight are not generally considered masterpieces.[2]

Perhaps vacations don't just encourage divine intervention and unconscious problem solutions. Maybe breaks from work help geniuses recover from exhaustion and allow them to forget the blind alleys they have been pursuing.

Another problem concerns procedures. Testing the *genius equals creativity equals novel-in-world-history* cliché calls for historical analyses that are unstable over time (fresh evidence can overturn who wins the title *first-in-the-history-of-the-world*) and notoriously controversial (Did Leibnitz or Newton discover calculus first, or did they both invent calculus simultaneously and independently in Germany and England? Nobody knows).

The third and most serious problem is: why should we care who wins the title *first-in-world-history*? The interesting and important question is: how does the human brain create ideas that are novel and useful? That's where Henry comes in. His inability to create new and valuable ideas helped reveal the brain mechanisms underlying the normal creativity inherent in every aspect of human life.

Everyday creativity is a basic biological given that shaped how our species evolved. Geniuses have driven that elementary ability to profound levels, but no solid evidence indicates that the brains of geniuses and ordinary folks operate in fundamentally different ways. Geniuses can create new and useful ideas because everyone can. The basic ability that allowed Shakespeare to create "Juliet is the sun" is available to anyone with a normal brain. Evolution built the capacity to generate metaphors into our genes.[3]

This recently elaborated concept of creativity counts any kind of new and useful idea as creative. New to *the brain that created it* but not necessarily new to every brain in recorded history. And useful to *someone*, not necessarily to humanity at large. From this perspective, even new and useful *personal insights* are creative. A father who helps his family by thinking up a new way to bribe his child into her pajamas is being creative. So is the husband who wins over his mother-in-law with a novel birthday present. So too the psychotherapist who helps a patient comprehend and solve a life-long personal problem.

This expanded view of creativity encompasses even everyday *comprehension*. Theater-goers are being creative if they form an internal representation of the concept *Juliet is the sun* that is *new* (not previously represented in their brain) and *useful* (for understanding Shakespeare's *Romeo and Juliet*). Although Benjamin Franklin clearly helped more people *in general* when he invented the bifocal lens, new comprehension that helps a single individual is also creative.

After reading that paragraph in an earlier draft, a colleague emailed, "Wait. Isn't the concept of creative comprehension too general? Doesn't it imply that *any* grammatically possible interpretation of an unfamiliar sentence is creative?" It does not. People often comprehend sentences in new and grammatical ways that are *uncreative* and *counterproductive* rather than useful in the real world. Here are three hypothetical examples: *Juliet is the sun* means that she is millions of miles from earth; *Juliet is the sun* means that the sun is named Juliet; and *Juliet is the sun* is silly because Juliet is human, and the sun is an object. All three interpretations are novel and grammatical, but *uncreative*. They *impair* comprehension of Shakespeare's play.

Now here is an actual example from my own personal experience as a child. My grandfather, a minister and spiritual leader in twentieth-century Nova Scotia, was delivering a sermon on the twenty-third Psalm, a passage that begins: "The Lord is my shepherd, I shall not want." When I heard this, I thought it was a misprint. At age five, I was certain that you should want the Lord your shepherd. I cannot remember anyone explicitly correcting my admittedly un-creative interpretation of "want." Somewhere along the winding road that leads to skepticism, I must have corrected that error myself.

Another, more common type of *non-creative* comprehension is *both* grammatical *and* useful. Every speaker of English has encountered this conversational exchange: "Good morning." *Listener response*: "Good morning." "How are you?" *Listener response*: "Fine." Despite its usefulness in promoting social bonding in the real world, this how-are-you dance is *routine*. Missing is the *surprise* that triggers formation of *new* internal representations inherent in all creativity. In fact, creative comprehension in that how-are-you exchange would be unwelcome. Out of the blue, most askers do not *actually* want to hear how you are.

Uncreative comprehension aside, many listener responses do display the novel and useful essence of creativity. Consider the creative comprehension in this excerpt from a conversation between a news reporter, Canadian prime minister Jean Chrétien, and US president Bill Clinton. Transcribed in 1997, the excerpt begins with Chrétien's response to the reporter's suggestion that his government investigate the 1996 increase in drugs entering the United States from Canada.[4]

Prime Minister Chrétien: It's more trade.
News Reporter: More drugs coming in from Canada to the United States?
President Bill Clinton: More **drugs**, she said. [emphasis in the original]
Prime Minister Chrétien: More drugs—I heard "trucks." [*Laughter*] I'm sorry.
President Clinton: I'm glad we clarified that, or otherwise he'd have to delay calling the [1997] election. [*Laughter*]

The president's *More **drugs**, she said* signals that Clinton had created two new internal representations, roughly, *Chrétien seems to believe that increased trade caused more drugs to enter the United States from Canada*, and *Chrétien probably misheard the question because a Canadian prime minister supporting enhanced drug traffic into the United States is unthinkable.*

By confirming that he did indeed mishear, Chrétien set the stage for Clinton's next contribution: *Without this clarification, Chrétien would have to repair the political fallout from his statement and postpone his call for an immediate Canadian election.* Clinton was genuinely *creative* here because the laughter his interpretation triggered was diplomatically helpful: Canadians soon re-elected Chrétien and relations between Canada and the United States were exceptionally warm and friendly for the next seven years.

If new and useful internal representations are the soul of creativity in comprehension and action, laboratory findings suggest that Henry's operation killed his creativity. However, direct understanding of creativity is impossible in the laboratory. Experiments are designed to examine the effects of one factor (here, hippocampal damage) while holding all other relevant factors constant (here, the age, education, IQ, skills, native language, socio-economic background of a comparison group). This makes experiments inherently unlike the real world—where a universe of factors (some unknown, many poorly understood) is free to vary and impossible to keep constant.[5] An essential aspect of creativity—real world value or usefulness—is beyond the reach of experiments. You will discover firmer evidence for the relation between creativity and the hippocampus in the next chapter, which describes the everyday conversational train wrecks in Henry's life.

QUESTIONS FOR REFLECTION: CHAPTER 12

1) Can you think of some creative ways to apply what you've learned in this chapter about creativity that might enhance your mental agility in everyday life?**
2) Before reading this chapter, did you equate creativity with genius-level ideas? How do you feel about creativity now?
3) How has your understanding of novelty and ambiguity changed after reading this chapter?
4) In what ways do you exhibit creativity in your everyday life?**
5) Do you think computers can be creative? How about nonhuman animals? Give some examples.

TEST YOUR MEMORY FOR CHAPTER 12

1) Why is equating creativity with genius a problem?
2) How is your real-world life quite unlike a laboratory experiment? Does this impact your ability to try to understand how your mind and the world around you works?**
3) Based on what you've learned from this chapter, how does your mind creatively combine novelty and usefulness?**

MEMORY TEST ANSWERS FOR CHAPTER 12

1) pp. 169–70
2) p. 173
3) p. 171

Chapter 13
THE MAN WHO MISTOOK A WASTEBASKET FOR A WINDOW

HOW DOES YOUR BRAIN KNOW WHEN SOMETHING IS WRONG?

Dr. Oliver Sacks, the famous author, physician, and professor of neurology, met Dr. P. in the mid-1960s. Decades later, Dr. Sacks characterized Dr. P. as an accomplished music teacher and singer, a man of great cultivation and charm who talked well and fluently, with imagination and humor. "What a lovely man," said Dr. Sacks to himself. He could not imagine that anything serious had brought Dr. P. to his clinic office.[1]

With Dr. P's wife by his side, Dr. Sacks conducted a routine neurological exam. Dr. P's coordination, reflexes, and muscle strength tested normal. Only after the exam did something truly bizarre happen.

As he prepared to leave, Dr. P. began to look around for his hat. He reached out his hand, grasped his wife's head and tried to lift it off and put it on his own head. He had mistaken his wife for a hat! His wife seemed unperturbed, as if accustomed to this curious behavior.

At the time, Dr. Sacks could make no sense of this incident. The term *face blindness* did not exist then and conventional neurology and neuropsychology lacked a way to explain how Dr. P. could, on the one hand, mistake his wife for a hat, and on the other, function as a distinguished teacher at a music school.

Around that same time, Dr. Brenda Milner discovered something equally puzzling.[2] She had just given Henry the Hidden-Figures test, a paper-and-pencil measure of perceptual ability. Henry's task was to find and trace an abstract target shape hidden within a complex drawing known as a concealing array. Figure 13.1 shows the target highlighted in the concealing array. A real-life analog might test a soldier's ability to detect an enemy combatant (the target) camouflaged in a forest (the concealing array).[3]

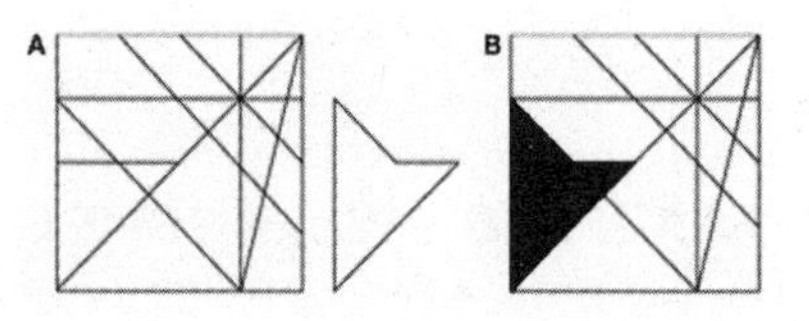

Figure 13.1. A Hidden-Figure revealed: the hidden-target at the right of Figure 13.1(a) is revealed in black within its concealing array in Figure 13.1(b). (Image in T. D. Ben-Soussan, A. Berkovich-Ohana, J. Glicksohn, and A. Goldstein, "A Suspended Act: Increased Reflectivity and Gender-Dependent Electrophysiological Change following Quadrato Motor Training," *Frontiers in Psychology* 5 [2014], https://www.frontiersin.org/articles/10.3389/fpsyg.2014.00055/full#B36.)

Henry found very few of the camouflaged targets. His poor performance on the Hidden-Figures test suggested a serious impairment in his ability to consciously recognize objects in complex visual scenes.[4]

This was curious. Henry should have aced the Hidden-Figures test. After all, he seemed to function without difficulty in his everyday visual world. He only appeared to experience problems in remembering where he had been or where he was going. Dr. Milner decided that Henry's abnormal Hidden-Figures score was unrepresentative of his true perceptual abilities. She ignored his Hidden-Figures deficit and moved on to other issues.

With Dr. Milner's considered opinion in mind, my UCLA lab decided to examine Henry's perceptual world in other ways. What we discovered was a man who mistook a wastebasket for a window. This and other spectacular errors in how Henry identified objects in a variety of perceptual tasks left a trail of evidence that eventually led back to his impaired performance in Milner's Hidden-Figures task.[5]

WHAT'S WRONG HERE?

School children love the game called *What's Wrong Here?*. It appears in popular books with titles such as *What's Wrong Here at School?*, *What's Wrong Here at the Amusement Park?*, and *What's Wrong Here: Hundreds of Zany Things to Find.*[6]

Every page of these books displays a complex everyday scene, for example, a school classroom containing over a hundred objects and busy people. Some of the objects are erroneous. They appear in anomalous or impossible contexts, say, a bird flying upside down in a fish bowl. Children enjoy discovering what's weird in what's-wrong-here pictures.

My lab converted the what's-wrong-here game into a test of Henry's perceptual skills. An experimenter handed Henry copies of what's-wrong-here pictures one at a time. Using a pencil, he circled as many erroneous objects as possible within a generous time limit and briefly explained what was wrong with each.

Henry correctly circled significantly fewer erroneous objects than the closely matched control participants who later took the same test. Henry also did things the controls never did. He circled and misidentified *normal* objects. An ordinary wastebasket on the floor beside a teacher's desk Henry called "a window" without correcting himself. A normally drawn rabbit in another picture Henry called "a dog," again without self-correction.[7]

Henry's *uncorrected object identification errors* were not slips of the tongue. People usually signal the occurrence of an error with an *um* or *er*, followed by a correction. Henry did none of that. Nor were Henry's uncorrected object identification errors due to poor visual acuity. He accurately identified letters presented in isolation as readily as controls of comparable age.

Subsequent experiments showed that Henry's perceptual errors were specific to scenes. He had difficulty disentangling visual objects from unfamiliar surroundings in complex displays. He readily identified windows, wastebaskets, dogs, rabbits, and mountains *when depicted in isolation*, but when the same objects appeared in real-world settings, Henry misidentified them.[8]

In one experiment, the task was to describe a drawing of executives at a board meeting discussing a business performance chart mounted on an easel. The chart displayed the ups-and-downs of corporate profits. Without hesitating, Henry called the chart "a window" and the peaks in profit "mountains in the distance."

In another experiment Henry experienced similar problems when asked to detect errors in what's-wrong-here sentences. Just as he could not identify erroneous objects in what's-wrong-here scenes, Henry could not identify erroneous words in ungrammatical sentences. He saw nothing wrong with *She hurt himself.* For Henry, ungrammatical sentences seemed error-free. His problem wasn't ignorance or impoverished vocabulary. Like the control participants, Henry had a high school degree. Henry could comprehend the familiar word *crush* in isolation but not in the unfamiliar context *She was crushed by his death*, just as he could recognize a familiar chart in isolation but not in the unfamiliar context of an executive board meeting.[9]

SURPRISE ME!

The normal brain quickly detects anomalous objects in three unconscious steps.[10] The first and most important step is to construct an accurate internal representation of the erroneous object, say, a door that is impossible to open because its hinges appear on the same side as its doorknob. Because participants have never previously encountered such a door, hippocampal engagement is necessary to accurately represent its novel features in the brain.

Next, participants must retrieve their internal representation for a normal, *un-erroneous* door. This retrieval process is easy. When we open doors over the course of our life, we retrieve and use our existing internal representation for a normal door thousands and perhaps millions of times.

A comparison process is the final step in anomaly detection: the brain must compare its internal representations of the normal versus dysfunctional door and register surprise when the two do not match.

I suspected that step one was the source of Henry's problems in detecting what's-wrong-here. By rendering new internal representations impossible to form, Henry's hippocampal damage would make erroneous doors indistinguishable from ordinary doors. For Henry, an impossible door would seem normal.

However, this hunch could be wrong. Henry's what's-wrong-here deficit could reflect the comparison processes in step three. The problem is that comparison processes and new memory formation are inseparable in the what's-wrong-here task. To test my hunch, I needed a new task. This led me to reconsider Brenda Milner's Hidden-Figures test as a possible candidate for separating out memory formation versus comparison processes.

HENRY'S HIDDEN-FIGURES PERFORMANCE REVISITED

I first re-examined Dr. Milner's concerns about the standard Hidden-Figures test as a measure of Henry's perceptual abilities. She pointed out sections of the test that required participants to search for *two* target figures in a lineup of concealing arrays. Perhaps Henry forgot one target while searching for the other. Milner attributed Henry's Hidden-Figures deficit to forgetting rather than to perception of the targets in their concealing arrays.[11]

To test Dr. Milner's forgetting hypothesis, I had my UCLA lab give Henry and suitable control participants Milner's Hidden-Figures test and *separately* analyze performance on the *typical* sections of the test that contained a single target figure resembling Figure 13.1(a). This had never been done before. We also added a new twist—a condition where the concealing arrays contained familiar targets such as a square, a circle, or a right-angle triangle, for comparison with the standard condition where the targets were unfamiliar, as in Figure 13.1(a).[12]

Because people detect familiar forms in everyday life using existing internal representations, the familiar targets should pop out of their concealing arrays as readily for Henry as for control participants if Henry's comparison processes are intact.

And that's what our results showed. Henry's problems were confined to *unfamiliar* target and concealed forms that had few or no prior encounters in everyday life. He correctly traced as many *familiar* targets in the concealing arrays as did the controls. Comparison wasn't Henry's problem. He easily compared target and embedded forms if both were familiar.

Forgetting one target while searching for the second in a two-target array was likewise not the sole basis for Henry's Hidden-Figures deficit. For both the one-target arrays resembling Figure 13.1(a) and for the two-target arrays that worried Dr. Milner, Henry correctly traced significantly fewer unfamiliar targets in the Hidden-Figures test than did the control participants.

Henry only experienced problems when target and concealed forms were *unfamiliar*, with few or no prior encounters in everyday life. He correctly traced significantly fewer unfamiliar targets in the Hidden-Figures test than did control participants, for both the one-target arrays resembling Figure 13.1(a) and for the two-target arrays that worried Dr. Milner. Forgetting one target while searching for the second in a two-target array could not explain Henry's Hidden-Figures deficit.

Although Dr. Milner was wrong in these respects, her general intuition was nonetheless correct. Henry had a memory problem. She just did not grasp that newly formed internal representations for perceiving forms in visual arrays are memories.[13]

PERCEPTUAL ERRORS IN THE HIDDEN-FIGURES TEST

We analyzed two extraordinary types of errors in our Hidden-Figures data: *target-unrelated* and *target-itself* errors. Together these errors revealed why unfamiliar targets were so difficult for Henry to detect in their concealing arrays: he could not create internal representations for novel forms.[14]

We scored *target-unrelated* errors when participants drew forms in a concealing array that bore little resemblance to the target. Figure 13.2 shows two examples. In Figure 13.2(a), Henry traced around the concealing array itself and drew a diagonal line within it, completely missing the target. In Figure 13.2(b), Henry traced a large diamond in the concealing array, failing to spot the target within it. No control participant made target-unrelated errors and Henry only made this type of error with unfamiliar targets.

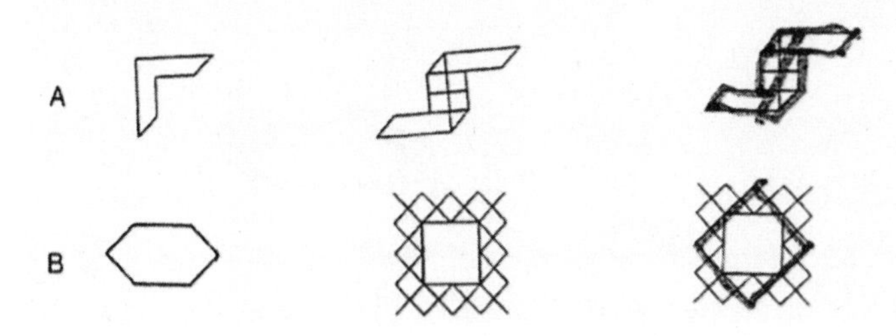

Figure 13.2. Examples of Henry's target-unrelated errors. Targets appear to the left, original concealing arrays appear in the middle, and Henry's errors appear in the concealing array to the right. (Image from D. G. MacKay and L. E. James, "Visual Cognition in Amnesic H.M.: Selective Deficits on the What's-Wrong-Here and Hidden-Figure Tasks," *Journal of Experimental and Clinical Neuropsychology* 31 [2009]: 769–89.)

Why were Henry's target-unrelated errors confined to unfamiliar targets? Unable to form internal representations for novel targets, tracing them in their concealing arrays was impossible. However, Henry avoided target-unrelated errors involving familiar targets by using existing internal representations to guide and constrain how he traced circles, triangles, and squares in their concealing arrays.[15]

We scored *target-itself* errors when participants drew lines in the target itself instead of in the concealing array. Like target-unrelated errors, only Henry made target-itself errors, and only with unfamiliar targets. For example, Henry extended existing lines and added new ones within the unfamiliar target in Figure 13.3, as if trying to see what the target might look like in a *provisional* concealing array that *he himself* created. Of course, lacking an internal representation for unfamiliar targets, this ploy did not work.[16]

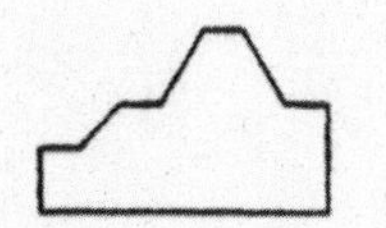

Figure 13.3. A target-itself error illustrated. The original target appears to the left with Henry's marked-up target to the right. No concealing array is shown. (Image from D. G. MacKay and L. E. James, "Visual Cognition in Amnesic H.M.: Selective Deficits on the What's-Wrong-Here and Hidden-Figure Tasks," *Journal of Experimental and Clinical Neuropsychology* 31 [2009]: 769–89.)

WHAT HAVE YOU DONE TO MY EYEGLASSES, DOCTOR?

Did Henry himself know that his hippocampal damage impaired his perceptual abilities? Henry was conscious during his operation and did not complain of pain. So what did Henry perceive as his neurosurgeon slowly removed his hippocampus? Nobody knows for sure. All we know is that a powerful association between brain surgery, his eyeglasses, and impaired vision remained entrenched in Henry's memory many decades after his operation. This three-way association surfaced in a story about his operation that Henry repeated with minor variations to anyone who would listen, sometimes three or four times a day. This is how Henry told the story to William Marslen-Wilson, as recorded and transcribed in 1970:

Henry: . . . uh . . . thank you . . . it's funny . . . the way . . . what I always wanted to be . . . though I know I couldn't . . . uh . . . well . . . put it down as the . . . well . . . why I couldn't be it . . . uh . . . naturally with the . . . epileptic seizures of any kind . . . but I thought of . . . the wearing the glasses.

Marslen-Wilson: Of . . . of what?

Henry: The wearing the glasses.

Marslen-Wilson: Yes . . .

Henry: . . . brain surgeon . . .

Marslen-Wilson: Yes . . . that's what you wanted to be? . . . oh, well . . .
Henry: . . . because I know, in brain surgery . . . that wearing glasses . . . these little bits [??] [makes gesture showing his hand slipping slightly] . . . that person is gone . . . that endured [*sic*].

Details of Henry's tale changed from one telling to the next. Sometimes the operating-room nurse accidently dislodged his glasses so that he couldn't focus. (It is extremely unlikely that Henry wore glasses during his surgery.) And sometimes little specks of blood or perhaps dirt spurted onto his glasses and impaired his ability to see. The only verifiable aspect of Henry's story is that he did possess eyeglasses not long after his surgery. Only the association between brain surgery, impaired vision, and eyeglasses remained constant in his story. And one of those aspects resurfaced again in the curious, well-documented and perhaps not unrelated incident where Henry threw his eyeglasses at his mother during a heated exchange.

Why his *eyeglasses*? And why *at his mother*? Of the many things that people throw at each other in anger, eyeglasses must come near the bottom of the list. Henry's action was more likely to damage his glasses than to hurt his mother. A police record shows that Henry did violently attack people with objects, but *eyeglasses* were not among them. So why did Henry attack his *mother* with *this* particular object rather than something that could do real harm? Perhaps the smudges on the eyeglasses in Henry's story were symbolic. Maybe Henry chose eyeglasses to throw at his mother because he was angry at her for allowing the brain surgery that forever impaired his perceptual abilities in ways that eyeglass could not fix. We'll never know for sure whether this is the correct analysis. Unlike Dr. Sacks, Henry was not seeing a psychiatrist.

QUESTIONS FOR REFLECTION: CHAPTER 13

1) In what way was the process of detecting anomalous objects in a visual scene and detecting ungrammatical words in a sentence parallel for Henry?
2) Henry struggled with identifying familiar objects in unfamiliar or

complex visual scenes, as in the what's-wrong-here test. The world around us is filled with familiar objects that can be difficult to identify in unusual or complex scenes. Have you ever encountered similar difficulties in experiencing the world around you? What are some ways you can train your brain to more easily and accurately identify objects in confusing contexts?**

3) After reading this chapter, what do you know about visual perception or language production that you did not know before?

TEST YOUR MEMORY FOR CHAPTER 13

1) If someone you know has difficulty, like Henry, with catching and correcting their ungrammatical errors when speaking, how can you use what you've learned in this chapter to explain what might be going on in their brain or mind?**
2) What are the three important steps for detecting anomalous or erroneous objects in a visual scene? Which step/s were most difficult for Henry?
3) Why would Henry have greater difficulty detecting two target features, versus only one, in a large visual array containing numerous distractors?

MEMORY TEST ANSWERS FOR CHAPTER 13

1) pp. 178–79
2) pp. 3–4
3) p. 180

Chapter 14
THE MYSTERIOUS FACE IN THE MIRROR

HOW CAN YOU ENHANCE YOUR CREATIVE PERCEPTION AND ACTION?

Henry was terrible at recognizing faces after his operation. If that had been his sole neurological problem, Henry's diagnosis might have been *prosopagnosia*, a scientific term that combines the Greek words *prosopon*, meaning *face*, with *agnosia*, meaning *not knowing.*[1] The translation into modern English is *face blindness.*

Face blindness is a serious problem for millions of otherwise normal people. One of them was Dr. Oliver Sacks, author of *The Man Who Mistook His Wife for a Hat*, the book discussed in the previous chapter. Throughout his life, recognizing faces was difficult for Dr. Sacks. A genetic defect was almost certainly the cause. At age seventy-seven, he still could not identify the faces of familiar acquaintances, including that of his psychiatrist.[2]

In one memorable story, Dr. Sacks had left his therapist's office just minutes earlier when a well-dressed stranger in the lobby of the building unexpectedly greeted him by name. This was puzzling. Dr. Sacks could not imagine who this audacious person might be.[3]

Then the doorman greeted the stranger by name—which Dr. Sacks immediately recognized as belonging to his analyst. Although Dr. Sacks could recognize familiar names, equally familiar faces were meaningless and unrecognizable. This embarrassing episode became a topic of discussion at their next session! The two men had spent hundreds of hours together over several years. His analyst found it easier to believe that Dr. Sacks failed to recognize him for psychological rather than genetic reasons.[4] What his psychiatrist did not realize was that Dr. Sacks could not even recognize *his own face* in novel or unexpected settings. On several occasions, Dr. Sacks apologized aloud for almost bumping into a large bearded man, only to discover that the stranger with the beard was his own reflection in a store window.[5]

Dr. Sacks later discovered the source of his problem: the brain contains more than thirty regions dedicated to processing different types of visual information. These specialized visual areas include independent modules for perceiving faces, motion, color, depth and geometric forms. The only impaired module in Dr. Sacks's brain was the one for perceiving faces. This *fusiform face module* consists of a network of intimately connected regions that includes parts of the lower occipital lobe and a subsection of the hippocampus in the medial temporal lobe (see Figure 14.1).

The hippocampal component of the face module becomes especially active when people process a face for the first time. Problems in recognizing and remembering faces virtually always ensue when genetic mutations prevent development of the fusiform face area, or when a stroke, tumor, infection, surgical intervention, or degenerative disease impairs its functioning.[6]

Typically, patients suffering from face blindness can see quite clearly the eyes, mouth, and other features of a face. They can even identify a close acquaintance via an unusual facial feature, say, unkempt red hair, bushy eyebrows, thick spectacles, or a prominent mole. Their problem is that the unique pattern of spatial relations linking eyes, nose, hair, and mouth eludes them. The features just do not seem to add up to a coherent face. The faces of Martin Luther King and President John F. Kennedy can seem indistinguishable despite the differences in skin color.

Repeatedly encountering a particular face sometimes reduces the problem, but not always. With genetic defects or major neurological damage, blindness for a particular face can persist despite massive exposure. This was the case with Dr. P., the patient Dr. Sacks described as "the man who mistook his wife for a hat." Unless his wife of many decades wore a distinctive ribbon at parties, Dr. P. could not find her when it came time to leave.[7]

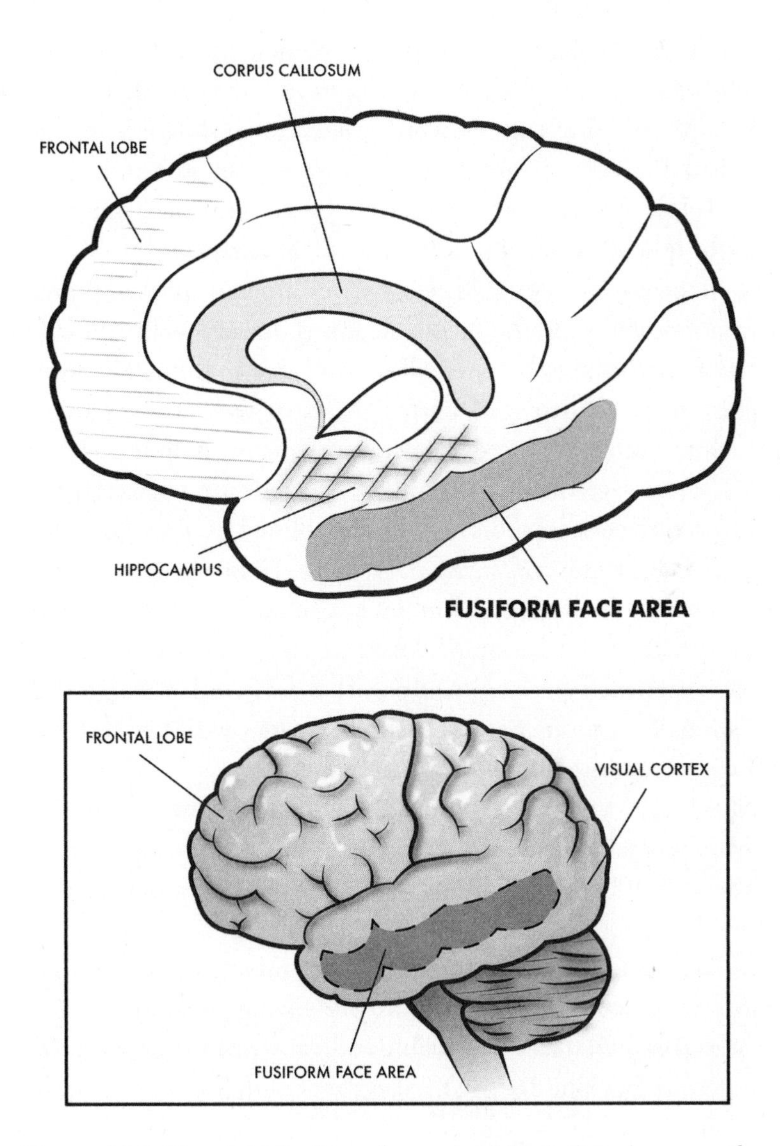

Figure 14.1. The right fusiform face area (shaded) is drawn as if the brain had been cut vertically down the middle (top panel) and as if one could see through the left cortex (bottom panel) to the right side. (Image by Patrick Phalen.)

NEW INSIGHTS INTO FACE BLINDNESS, *AND HENRY*

Similarities between the normal and abnormal aspects of Henry's language and visual cognition suggested some new ideas about face blindness, *and about Henry*. In the case of language, Henry's ability to comprehend the meanings of familiar words presented in isolation was normal, but he had difficulty integrating those word meanings into their context in a novel sentence. That integration process required hippocampal drivers for quickly forming new internal representations.[8]

Nonetheless, Henry could *gradually* create new internal representations via repeated exposure or use. *Repeated activation*, the most fundamental of all learning processes, allowed Henry to form and strengthen synapses in his undamaged cortex. Massive reencounters eventually helped him recognize his own face in a mirror, despite its changing appearance as he gradually aged and became seriously overweight.

So too with patients suffering from face blindness. Prosopagnosics can recognize facial features *depicted alone*, just as Henry understood familiar words in isolation. Their problem lies in combining facial features into a novel internal representation of the entire face. Similarly, many prosopagnosics experience special difficulty with faces in unfamiliar spatial contexts, just as Henry experienced special difficulty with familiar words in novel sentence contexts.

Likewise, just as daily exposure helped Henry slowly learn to recognize his own face in the mirror, so too with most prosopagnosics (Dr. Sacks and Dr. P. being exceptions because of their extensive neurological damage).[9]

To me, those similarities suggested that face blindness and Henry's memory problems reflect analogous underlying processes. The difference was that *only* faces are problematic for prosopagnosics. Despite his profound face blindness, Dr. Sacks was perhaps more famous than any other MD in history for his ability to combine words into exceptionally clear and coherent sentences. One reviewer said it all: "Damn, the man can write."[10] His genetic defect spared the drivers in his brain for quickly binding words into novel internal representations. Only his drivers for swiftly creating novel internal representations for faces were dysfunctional.

The idea that brain damage could selectively spare specific types of drivers raised an interesting question in my research with Henry. Henry's operation spared about half of his hippocampus. What if one of his hippocampal drivers remained functional? This would allow Henry to quickly form new internal representations for some specific category of information. All I needed to do was discover what specific category that might be. A later chapter shows how this line of reasoning paid off: just as Dr. Sacks's genetic defect spared his hippocampal drivers for rapidly forming *any* new linguistic unit, Henry's operation spared at least two of *his* hippocampal mechanisms for binding linguistic information.

NEW INSIGHTS INTO VISUAL COGNITION

Meanwhile, I had some new research ideas about visual cognition that I wanted to explore. One was a possible connection between hippocampal drivers and consciousness that I hoped to test using experimental procedures developed by Dr. Anne Treisman, a distinguished psychologist at Princeton University. Her experiments showed that consciousness comes in two flavors: *automatic* versus *sought-for*. The *automatic* type of consciousness is effortless and quick: objects just "pop into perceptual awareness." The *sought-for* type of consciousness is arduous and slow. My hypothesis was that achieving *sought-for* consciousness of an object reflects the formation of new internal representations by the hippocampus.

Because Dr. Treisman's procedures measure how quickly *anyone* becomes aware of *any* object, including the simple, familiar form of a letter, let's try to replicate her results for a bold **R** immersed in a sea of normal *R*s. Please point at the bold **R** in the array below as quickly as possible.

RRRRRRRRRRRRRRR
RRRRRRRRRRRRRRR
RRRRRRRRRRRRRRR
RRRRRRRRRRRRRRR
R**R**RRRRRRRRRRRRR
RRRRRRRRRRRRRRR
RRRRRRRRRRRRRRR
RRRRRRRRRRRRRRR
RRRRRRRRRRRRRRR
RRRRRRRRRRRRRRR
RRRRRRRRRRRRRRR
RRRRRRRRRRRRRRR

You probably became aware of that bold **R** very quickly, just like Dr. Treisman's undergraduates. Also like the students, you would have discovered that bold **R** equally quickly if the array had contained more letters in longer rows.

Now ask yourself this: did the bold **R** come effortlessly into consciousness without a letter-by-letter search through each row in the array? If you responded *yes*, then you experienced Dr. Treisman's *automatic* form of awareness, where an object just "pops into consciousness" quickly and without effort.[11]

How did you so easily become aware of that bold **R**? Under my hypothesis, only the color black in the sea of letters entered your awareness and guided your pointing. To respond, you did not have to internally represent the *combination* of form and color for any of the letters in the array, just the color black. Henry should experience Treisman's *automatic* pop-up awareness without difficulty.

Now let's try to replicate Dr. Treisman's results for a bold **R** immersed in a different sea of letters. Please point at the bold **R** in the array below as quickly as possible.

P R **P** R **P** R P R R **P** R P R **P**

P P R **P** R P R **P** R P R **P** R P

R **P** R P R **P** R P R **P** R P **P** R

P R **P** R **P** R P R P R **P** R R **P**

P R P R P **P** R P **P** R P R **P** R

R R **P** R P R **P** **P** P R **P** R P R

P **P** R P R R P **P** **P** R P R P **P**

R **P** **P** R P R **P** R P R **P** R P R

P R P **P** R P R P **P** R R **P** **P** R

R **P** P R **P** P **P** R **P** P R **P** R P

R **P** **P** P R **P** **R** R **P** R P R P R

R **P** R P R P **P** R P R **P** R **P** P

Like Dr. Treisman's participants, if you found that bold **R** at all, it probably took a long time (it's in the second to last row, seven letters from the left). Also like her students, you would have required much more time to discover the bold **R** if the array had contained more and longer rows of letters.

Now ask yourself the same question as before: did the bold **R** come effortlessly into awareness without a row-by-row, letter-by-letter search through the array? If you responded *no*, you experienced Dr. Treisman's *sought-for* type of conscious perception.[12]

Why was becoming aware of the bold **R** so difficult in this case? It wasn't just because the bold **R** was near the bottom in this example. Dr. Treisman randomized the positions of the targets. My hypothesis was that time and effort was needed to search through the sea of forms to find the one that was both an **R** and bold. You had to create new internal representations of the form-color pairing for most of the letters in the array in order to find the combination that matched the target.

What brain mechanism would do the combining of form and color in my theory? Hippocampal drivers![13] Because of Henry's hippocampal damage, Dr. Treisman's *sought-for* awareness should elude him. Unable to form new internal representations of the form-color combination for any of the letters, Henry should never find the bold **R** in this condition, regardless of how long or hard he tried. However, Henry should easily find the bold **R** in the sea of normal Rs.

Alas, I was not able to test these predictions. I no longer had permission to test Henry. However, this new situation also had an upside. It gave me time to explore some interesting relations between Henry's visual binding errors and the binding errors that normal people make when processing visual forms and sentences under time pressure.

NOW YOU SEE VISUAL BINDING ERRORS, NOW YOU DON'T

Normal individuals sometimes misperceive visual forms in ways that resemble Henry's visual binding errors. Again it was Dr. Anne Treisman who discovered this. She asked undergraduates to say what they saw after she briefly flashed a visual display on a computer monitor. Each display resembled Figure 14.2, except that the number at each side appeared in green, and the three forms in between appeared in different colors rather than in shades of grey. Participants were instructed to first report the numbers, then the color and shape of the three forms.

Figure 14.2. A typical visual display for demonstrating illusory conjunctions. The numbers were presented in green, the triangle in orange, the oval in blue, and the circle in black. (Image by Micah Johnson, adapted from A. Treisman and G. Gelade, "A Feature-Integration Theory of Attention," *Cognitive Psychology* 12 [1980]: 97–136.)

If the presentation times were just right, participants reported the correct numbers, colors and shapes, but they mixed up what color went with what shape. For the display in Figure 14.2, they might report that they saw a blue triangle when the triangle was in fact orange, or a black oval when the oval was in fact blue, or an orange circle when the circle was in fact black. However, they were not just guessing. They were absolutely certain those shapes appeared in those colors.[14]

Dr. Treisman called these perceptual errors *illusory conjunctions.* She found them interesting because they were clearly occurring at some level well above the receptors in the retina. Participants correctly perceived the forms and colors, but *combined* them incorrectly when forming internal representations of the visual objects in the display.[15]

What makes illusory conjunctions even more interesting is their relation to the hippocampus and Henry's visual binding errors: participants only experienced illusory conjunctions when they had to form *new* internal representations, not when they could retrieve familiar internal representations from memory. In the experiment that showed this, Dr. Treisman informed one group of participants that they would see displays of abstract forms painted in specific colors. If Figure 14.2 was the upcoming display, these participants were told to expect a blue oval, an orange triangle, and a black circle. To accurately perceive and report these abstract visual objects in the correct order, shape, and color, participants in this *abstract-form* group had to create new internal representations.

Dr. Treisman then presented the same displays for the same brief exposure times to a second group of participants. The only difference was that these participants expected each display to contain familiar but abstractly depicted *objects*. If Figure 14.2 was the upcoming display, participants in this *familiar-object* group were told to expect a blue lake, an orange carrot, and a black tire. They did not have to create new internal representations of the form-color combination for each object. They could identify the color of the objects by retrieving internal representations formed during childhood for recognizing carrots, lakes, and tires.

Results for the two groups differed greatly. Participants in the *abstract-form* group reported large numbers of illusory conjunctions. Because of the brief exposure times, they often perceived the expected forms in the wrong color. However, participants in the *familiar-object* group reported no illusory conjunctions whatsoever. Their perception of the expected color-form combinations was always accurate: the "carrot" always seemed orange, the "lake" always seemed blue, and the "tire" always seemed black.

This was not because the *familiar-object* participants just perceived whatever they expected to see. Intermixed among the *expected* displays presented to both groups were *unexpected* displays. When the abstract-form participants expected to see a blue oval, an orange triangle, and a black circle, their *unexpected* display showed a black triangle, an orange oval, and a blue circle. And when the familiar-object participants expected to see a blue lake, an orange carrot, and a black tire, their *unexpected* display showed a black carrot, an orange lake, and a blue tire. For these *unexpected* displays, the familiar-object group reported as many illusory conjunctions as the abstract-form group, and no more illusory conjunctions than the abstract-form group based on what they expected to see. Participants generated illusory conjunctions when they had to form new internal representations for unexpected shape-color combinations but not when they could base their perception on internal shape-color representations formed during childhood. As with Henry's visual binding errors, the critical determinant of illusory conjunctions was the hippocampus and the formation of new internal representations.

NEW INSIGHTS INTO SENTENCE COMPREHENSION

Being unable to test Henry also gave me time to explore the interplay of new versus old internal representations during the normal comprehension of sentences. Like Henry, ordinary folks can make spectacular errors when comprehending sentences. Perhaps I could apply what I now knew about Henry to help unravel how the normal brain understands language.

I was especially interested in how normal individuals answer comprehension questions such as *How many animals of each kind did Moses take on the ark?* Even though they know that Noah, not Moses, launched the ark, they usually respond *two.* However, that is wrong. The correct answer is *zero* because Moses did not take animals on an ark. Why doesn't knowledge about Noah and Moses stored in our brain prevent this type of comprehension error? Why do people think they correctly understand Moses questions when they do not?[16]

Another aspect of Moses mistakes interested me. Like Henry, normal folks in a Moses experiment firmly believe that their mistakes are correct. They never say, *Moses took two, I mean, zero animals on the ark,* spontaneously correcting their comprehension errors the way they routinely correct everyday speech errors.

It's not that participants answer too quickly or do not expect questions like this. When the experimenter illustrates this type of trick question in advance and warns against hasty or impulsive responding, folks with unlimited time to respond still make Moses mistakes. It is also not the case that a momentary lapse of attention causes participants to overlook the word *Moses.* Participants can correctly read the question aloud before responding and still make large numbers of Moses mistakes.[17] This fascinated me. It was as if one part of the brain accurately registers and reads aloud the word *Moses* while another part of the brain responds as if the question were about *Noah.* The participants must be comprehending Moses to mean Noah.

To test this idea, I asked Meredith Shafto, a UCLA graduate student at the time, to combine a Moses experiment with a surprise memory test for the gist of the Moses questions. If participants incorrectly responded *two* to the original Moses question and later remembered that the question was about Noah, they must have miscomprehended *Moses* as *Noah.*[18]

And that was what the data showed. After participants read aloud and answered the original questions in the Moses experiment, they received a list of *memory questions* about those original questions. For each memory question, they indicated whether the question was identical in meaning to one of the original questions that they had answered. They were to ignore changes in word order that did not alter what the original question meant. For example, if participants had originally answered the question *How many animals of each kind did Noah take on the ark?*, they should respond *yes* for the memory question *Noah took how many animals of each kind on the ark?* and *no* for the original questions.

What the participants did not know was that the correct response to *every* memory question was *no*, because none of the memory questions were synonymous with the original question. By responding *yes*, participants were unintentionally revealing that they had miscomprehended *Moses* as *Noah* in the original question.

The results showed that participants always recognized that the original versus memory questions differed in meaning if they had correctly responded *zero* in the main experiment. They knew that the original Moses questions were about Moses not Noah. However, if participants answered *two* to the original question, they always responded *yes* to the corresponding memory question. They thought the Moses questions were about Noah, even though they had accurately read the word *Moses* aloud before responding.

This finding explains why most participants believe that their incorrect response is correct when they make Moses mistakes. For them, *their response was in fact correct* because they misunderstood Moses to mean Noah. This also explains why they virtually never spontaneously corrected their Moses mistakes. Error correction only becomes possible when participants form a new and accurate internal representation of the question. They have no basis for detecting and correcting their Moses mistakes when their internal representation of the question fits their response.

Why do participants comprehend and respond as if Moses questions are about *Noah* even after reading the word *Moses* aloud? People often think one thing and say something different. This what happens *routinely* when we make speech errors. We say things that do not correspond to the idea we want to express. Correcting such speech errors is easy. As speakers, we normally create an internal representation of what we intend to say before we say it. This allows us to quickly detect and correct the mismatch between our spoken error versus our intended utterance. However, correction is difficult in comprehension. Comprehenders can utter the syllables in *Moses*, while at the same time constructing the incorrect internal representation, *This question is about how many animals* Noah *took on the ark.* Without help from outside, this mistaken comprehension is uncorrectable.

My studies of Moses mistakes demonstrated that the normal brain engages two fundamentally different types of comprehension processes. When people respond *zero* to questions like *How many animals of each kind did Moses take on the ark?*, their comprehension process is *creative.* The hippocampal region of their brain has formed an internal representation of the question that is new, accurate, and useful. However, when participants respond *two* to Moses-like questions, their comprehension process is *routine.* Instead of comprehending the question, they activate an existing memory representation—in this example, the story they learned as children about Noah and the ark. Subsequent chapters will show that routine comprehension was the only option open to Henry. His hippocampal damage rendered creative comprehension impossible. Had I been able to ask Henry some Moses questions, he almost certainly would not have responded creatively. He could replay the story of Moses and the ark, but he could not create a new and useful internal representation that integrated the unexpected word *Moses* into the context *How many animals of each kind were taken on the ark?*

REFLECTION BOX 14.1: CREATIVITY, NORMAL AGING, AND THE HIPPOCAMPUS

Your ability to create new internal representations in comprehension and action is remarkably immune to effects of aging. For ordinary older adults, creating new and useful plans and insights late in life is common. Even genius-level creativity can sometimes continue unabated into old age. At age ninety-one, Frank Lloyd Wright designed and completed the Guggenheim Museum, an architectural masterpiece. At age eighty, Giuseppe Verdi wrote one of his best operas (*Falstaff*). At age seventy-eight, Benjamin Franklin invented the bifocal lens, a creative feat welcomed throughout the world.

The phenomenon of neurogenesis, the biological process that forms new neurons, probably holds the key that lets you, me and other humans (including geniuses) create new and useful ideas as we age. The hippocampus provides the basis for creativity, and unlike many other brain regions, it continues to give birth to new neurons throughout adulthood and into old age.[19]

QUESTIONS FOR REFLECTION: CHAPTER 14

1) Many of us have experienced a temporary version of prosopagnosia, or face blindness, when encountering a familiar face in an unexpected location. For example, if you only see your cashier at the grocery store or the news reporter on the TV, it may be very difficult to quickly identify that person if you see them at your local gym. What have you learned in this and the prior chapter that might help to explain why this happens? Can you think of any ways to reduce this difficulty the next time you see someone in an unexpected context?
2) What are some examples of creativity in comprehension or action in your own life? What are some ways you can continue to enhance your creative comprehension and action?**
3) Most people easily recognize and distinguish between human faces but not between faces of animals. Have you experienced this *cross-species prosopagnosia*, perhaps when you mistook a neighbor's puppy or kitten as your own? Why do think this happens? Do you think you could get better at recognizing the faces of animals if you practiced?
4) Consider the automatic "pop-up" effect that Dr. Treisman demonstrated. Can you think of similar examples in everyday life? How could you use this pop-up effect to make it easier to find things, such as your favorite paragraph in a book, or your phone or car keys, or a favorite food item in an overstuffed fridge?**
5) Why do you think the human brain has developed the "two flavors" of consciousness (automatic vs sought-for)?
6) Figure 14.2 illustrated the illusory conjunction phenomenon

demonstrated by Dr. Anne Treisman. Why do you think that her participants were so convinced that they were right, when they were not?**

7) Have you experienced the Moses illusion? How might you be able to avoid making this mistake?

TEST YOUR MEMORY FOR CHAPTER 14

1) People with prosopagnosia have extreme difficulty recognizing faces, despite being able to see clearly the nose, eyes, and mouth, and to identify faces with unusual features. What is the fundamental basis of this disorder and how does it relate to Henry's problems with language and vision?
2) Many people with prosopagnosia can eventually learn to recognize their own or others' faces. What allows them to do this?
3) Compare and contrast the "two flavors" of consciousness identified by the famous psychologist, Dr. Anne Treisman. Can you think of some examples in your own life?
4) Where is the area of the brain that processes faces?
5) What is the important parallel between the patient Dr. P's face blindness and Henry's difficulties with language and visual perception?
6) How do illusory conjunctions in visual perception relate to Henry's memory and language problems? Can you think of some examples in your own life?**
7) Experiments on the Moses illusion show that people who wrongly answer the question, *How many animals of each kind did Moses take on the ark?* almost never correct themselves. Why not?
8) What two types of comprehension process play a role in the Moses illusion? Which type of comprehension process was intact in Henry?
9) What makes the hippocampus special in the brain in terms of longevity of neuronal function?

10) What famous neurologist treated the man who thought his wife was a hat? What famous psychologist first demonstrated Henry's difficulties with visual perception?

MEMORY TEST ANSWERS FOR CHAPTER 14

1) p. 186
2) p. 188
3) pp. 189–90
4) p. 186
5) pp. 188–89
6) p. 193
7) p. 196
8) pp. 196–97
9) Reflection Box 14.1, pp. 197–98
10) p. 185

Chapter 15
METAPHORS BE WITH YOU, HENRY

CAN A WHOLESOME DIET OF METAPHORS HELP NURTURE YOUR BRAIN?

Metaphors are more than enjoyable—the stuff of poetry and great literature. Without our knowing it, metaphors influence how we think about everyday ideas. Linguist George Lakoff and colleagues have documented hundreds of metaphors that people seem to live by.[1] One is the military metaphor that shapes how many people think about verbal disagreements. We consider arguments life-or-death battles that can be won or lost. As in real wars and battles, we think about gaining or losing ground in an argument, taking and defending a position, attacking *weak points*, abandoning *indefensible positions*, adopting new *lines of attack*, *demolishing*, *wiping out* or *shooting down* the ideas of an opponent, and launching critiques that are *right on target.*

Of course, we can also conceptualize the conflicting ideas underlying personal disagreements in other ways. We can liken them to people, plants, products, commodities, money, buildings, tools, and fashions. We can consider them *food for thought*, *difficult to swallow*, *hard to digest*, *half-baked*, *warmed-over and stale*, imparting *a bad taste in the mouth*.[2] These contrasting viewpoints are also metaphors and an important ongoing research question is which metaphors genuinely dominate our thinking and why. Adults may no longer imagine mountains having feet or think of breakfast as breaking a fast, but some of us still view life as but a dream.

Besides shaping everyday thought, metaphors represent an important educational tool —a way for familiar concepts to help communicate new ideas that are complex, abstract or otherwise difficult to learn and understand. Conceptual likeness is the key that opens this educational toolbox. Abstract resemblances between antibodies and a lock and key help convey how the immune system functions. Conceptual resemblances between DNA and an encrypted verbal message help communicate how genes work. Theoretical links between atoms and objects that are bouncy, sticky, and extremely tiny help people understand chemical reactions.[3] Abstract similarities between Juliet and the sun helped Shakespeare's audiences understand Romeo's surprising new relationship with Juliet.

SEX, METAPHORS, AND IDEAS ABOUT PEOPLE

The dual functions of metaphor—educating and persuading—overlap in personification, the most common type of metaphor in world literature. For example, classic personification metaphors that compare ideas and objects to male and female persons simultaneously educate children about sexual stereotypes and unconsciously shape aspects of their thinking for the remainder of their life. A wide range of scientific data indicate this, including the PhD thesis of Dr. Toshi Konishi, my former graduate student. Dr. Konishi first analyzed what concepts got personified as female versus male in an enormous sample of gendered personifications in literature written for English-speaking children, for instance, *Mother Nature* and *Father Time.* Her results were remarkably consistent. The concepts *nature* and *old age* always got personified female. So did *ships*, *countries*, *cars*, *the moon*, and relatively small and weak animals such as *mice* and *cats.* However, *time*, *death*, and *the sun* always got personified male, along with relatively large and strong animals such as *dogs* and *lions*.[4]

Dr. Konishi next examined how adults thought about male versus female people (e.g., *uncle* versus *aunt*) and concepts consistently personified male versus female in children's literature (e.g., *lions* versus *mice*). She intermixed both types of words and gave them to UCLA undergraduates to rate on dimensions such as warm-cold, likeable-unlikeable, powerful-powerless, and active-passive. The rating results for the two types of words exhibited the same pattern. The students rated both female terms (e.g., *sister*, *aunt*, *mother*) and concepts consistently personified *female* in children's literature as warm, likeable, and somewhat passive. However, they rated male terms (*brother*, *uncle*, *father*) and concepts consistently personified *male* in children's literature as powerful and somewhat cold and unlikeable.

Dr. Konishi then turned to languages such as Spanish and German that mark nouns as either masculine or feminine. Her hypothesis was twofold: that masculine gender markers (e.g., *el* and *der*) represent a form of male personification signaled by use of the pronoun designating human males (translated *he* in English), and that feminine gender markers (e.g., *la* and *die*) represent a form of female personification signaled by use of the pronoun designating human females (translated *she* in English). To test this hypothesis, Dr. Konishi asked native speakers of Spanish and German to rate a set of identical concepts (e.g., *sol* and *Sonne* without their gender markers) on dimensions such as warm-cold, likeable-unlikeable, powerful-powerless, and active-passive (translated into Spanish or German). Her choice of concepts was strategic. Unbeknownst to the participants, concepts marked masculine in Spanish (e.g., *el sol*, meaning *the sun*) were feminine in German (e.g., *die Sonne*, meaning *the sun*) and vice versa: concepts marked feminine in Spanish (e.g., *la luna*, meaning *the moon*) were masculine in German (e.g., *der Mond*, meaning *the moon*). Her rating results contradicted the centuries-old assumption that the obligatory gender-markers in these language carry no psychological significance. Spanish speakers rated concepts marked male in Spanish as like the stereotypical male person in some respects (e.g., powerful), and rated concepts marked female in Spanish as like the stereotypical female person in some respects (e.g., weak). In the unconscious minds of Spanish speakers, the sun is in control, whereas the moon is passive and subservient—a weak reflection of the powerful sun. When expressed in German, however, the identical concepts exhibited the opposite pattern of ratings: German speakers rated the concept *moon* (marked male in German) as like stereotypical male person in some respects (e.g., active, mildly unlikeable), and rated the concept *sun* (marked female in German) as like the stereotypical female person in some respects (e.g., weak, likeable).[5] Unconsciously, the German and Spanish speakers were processing the reversed gender markers for those concepts in the same metaphoric manner: male gender-markers called up stereotypes about human males and female gender-markers called up stereotypes about human females.

Completing the picture, Dr. Konishi returned to the concepts consistently personified female versus male in literature written for English-speaking children. She had UCLA undergraduates rate those concepts on the same dimensions as her German and Spanish speakers. Again the corresponding male versus female stereotypes emerged. Concepts consistently personified female in English (e.g., *old age*) were rated as passive and weak, like the stereotypical female person, but concepts consistently personified male (e.g., *time*) were rated as active, powerful and domineering, like the stereotypical male person. In the unconscious minds of English speakers, *time* is in control, and *old age* is the docile, obedient outcome. Konishi concluded that person metaphors allow people to transfer habitual attitudes from their social world to the world of ideas and things, and thereby relate in a familiar and personal way to newly encountered concepts and objects. In so doing, personification both perpetuates traditional gender stereotypes and teaches culture-specific attitudes in extremely subtle ways.[6]

GOT METAPHORS, HENRY?

Could Henry enjoy the educational and conceptual benefits of metaphors? Did metaphors influence Henry's thoughts and help him understand and learn? Was Henry able to express himself metaphorically? Could Henry use abstract conceptual likenesses as a tool for forming new internal representations?

A frequently repeated anecdote suggests that Henry could indeed understand and generate fresh metaphors. The story goes like this. Dr. Suzanne Corkin encountered Henry working on a crossword puzzle and commented, "Henry, you're the puzzle king." To which Henry replied, "Yes, I'm puzzling," as if he fully understood Corkin's novel *puzzle king* metaphor and wanted to advance the conversation with a witty double reference to his crossword puzzle habit and profound amnesia—an existential condition that puzzled him even in old age.

Does this anecdote really show that Henry comprehended Corkin's *puzzle king* metaphor to mean *You are a king among solvers of crossword puzzles*? Being puzzling has nothing to do with dominating at solving puzzles. Henry's *I'm puzzling* may have been just another of his incomprehensible non-sequiturs.

Another possible interpretation goes like this: focusing on his puzzle, Henry *misheard* Corkin to say *puzzling* rather than *puzzle king*. This being the case, *I'm puzzling* represents Henry's way of saying, *Yes, I am working on a puzzle*—which makes perfect sense in this context and is consistent with Henry's sentence comprehension deficits in carefully controlled experimental studies. The hypothesis that Henry can comprehend fresh metaphors demands better data.

When I still had permission to test Henry, I decided to determine whether Henry could understand the novel metaphors on the standardized Test of Language Competence (TLC).[7] On each trial Henry and closely matched memory-normal participants saw a short, metaphoric sentence, for example, *Maybe we should stew over his suggestion*, together with three possible interpretations of that sentence, in this case, *Let's think about it some more*, *Maybe we should put more meat into his idea*, and *Let's make sure to cook the stew long enough*. Their task was to choose the best interpretation, here, *Let's think about it some more.*[8]

The results indicated that Henry could not comprehend metaphors. He chose the correct interpretation reliably less often than normal individuals his age. He could have performed equally well by letting a coin toss determine his choices.

However, *which* incorrect interpretation Henry chose was informative. Neither incorrect alternative captured what the target sentence meant, but one always contained a word from the target sentence, for example, the word *stew* in *Let's make sure to cook the stew long enough*. The results indicated that Henry usually selected *that* wrong interpretation rather than the one without overlapping words, in this example, *Maybe we should put more meat into his idea.*[9]

This curious preference indicates that Henry understood neither the target sentences nor their possible interpretations *as sentences*! He simply chose the alternative with the overlapping word, *regardless of what the overall sentence meant*. Once again, our results showed that Henry can understand the meaning of familiar words in isolation but cannot integrate those word meanings into their novel contexts in sentences. That integration process requires the formation of novel internal representations—a creative action that yields different interpretations for *stew* in the contexts *cooking a stew* versus *stewing over a suggestion.*

A functioning hippocampus is essential for integrating meanings in that way. Just as Henry could understand the meaning of ambiguous words in isolation, but not in their novel context in a sentence, he understood the words in metaphoric sentences without comprehending the overall meaning of the sentence. Understanding familiar words requires only routine retrieval processes that do not engage the hippocampus. However, hippocampal mechanisms are needed for creating the new internal representations required to comprehend one kind of event—*taking the time to talk and think about something*—in terms of another—*slowly cooking, as with a stew* in sentences such as *Maybe we should stew over his suggestion.*

REFLECTION BOX 15.2: METAPHORS BE WITH YOU AND YOUR HIPPOCAMPUS

New neurons are like babies. They need stimulation and challenges. The hippocampus continues to give birth to new neurons throughout life, but without stimulation and challenges, those newborn hippocampal cells die.[10] Exercising your creativity is exactly the type of workout they need. My own experience writing this book can perhaps illustrate. To write a general audience book like *Remembering*, I needed to recognize and generate creative metaphors. Not just concrete, literary metaphors like Shakespeare's "Juliet is the sun," but *abstract* metaphors for explaining psychological and biological concepts to people who may not be neuroscientists. For me, this meant overcoming a powerful habit. I had for fifty years written books and articles in a standard scientific style that eschews metaphors. To break that habit, I played a find-the-metaphor game with every book for general audiences that I subsequently read while writing *Remembering*. Finding the metaphors was easy and fun. They were everywhere. The fifteen highlighted in the three sentences below appeared in a beautifully written book about the mind:[11]

Music **borrows** some of its mental **machinery** from language—in particular from prosody, the **contours** of sound that **span** many syllables. The metrical structure of **strong** and **weak** beats, the intonation **contour** of **rising** and **falling** pitch, and the hierarchical grouping of phrases all **work** in similar ways in language and in music. The parallel may account for the **gut feeling** that a musical **piece conveys** a complex message, that it **makes assertions** by introducing topics and **commenting** on them, and that it emphasizes some portions and **whispers others as asides**.

Playing this find-the-metaphor game exercised my hippocampus, improved my writing style and may have helped some newly formed hippocampal neurons to survive.

QUESTIONS FOR REFLECTION: CHAPTER 15

1) This chapter discusses common metaphors in our daily life, for example, the military metaphor for thinking about verbal disagreements. Can you think of some other examples of metaphor categories in your conversational speech? Do you think such language metaphors influence how you think or act?**
2) This chapter discusses different conceptual, abstract, or theoretical resemblances making up metaphors and how they can facilitate learning and understanding. Can you think of some additional examples from your own life? How might metaphoric thinking sometimes impair learning and understanding?**

3) How might a wholesome diet of metaphors be nutritious for the health of your brain?**
4) Try playing the find-the-metaphor game in what you read or write. Does this game help you think about language in new and interesting ways?**

TEST YOUR MEMORY FOR CHAPTER 15

1) What is gender personification and does it really differ across languages in its effects on thought? Do you think that this phenomenon influences the way you, or others you know, think about the world?**
2) How might the personification metaphor influence how you interact with the world?

MEMORY TEST ANSWERS FOR CHAPTER 15

1) pp. 202–203
2) pp. 202–203

Chapter 16
WHAT'S IN A NAME, HENRY?

WHICH ARE YOU, AN INTUITIVE LINGUIST OR AN INTUITIVE PSYCHOLOGIST?

A shadow haunted Henry after his operation. It followed him everywhere. It was the name *amnesic.* Until the *New York Times* published his real name in 2008, the world only knew Henry as "amnesic H.M."

Names are important. In the spring of 1966, the label *amnesic* could have cost me my PhD. Here's what happened. At the suggestion of my mentor and founding chair of the MIT Psychology Department, Dr. Hans-Lukas Teuber, I tested Henry's sentence comprehension. Without asking how Henry did on my test, Dr. Teuber invited me to present my results at a lab meeting the next day. In attendance was an international group of neuroscientists that included a distinguished visitor from McGill University, Dr. Brenda Milner, and Dr. Suzanne Corkin, who was Dr. Milner's former graduate student and Dr. Teuber's post-doctoral fellow.

Unbeknownst to me, everyone in attendance considered Henry a *pure amnesic* as defined in the *Diagnostic and Statistical Manual of Mental Disorders* (*DSM*). That standard handbook for clinical psychiatrists, psychologists and neurologists has been revised since then, but the current version (*DSM-V*) still defines amnesia as a severe inability to learn or remember new facts and events, with no chronic cognitive impairments such as dementia that "affect a patient's social and occupational functioning" and no short-term cause such as a brief bout of delirium, substance abuse or withdrawal.

For my audience that day, Henry's poor performance on my sentence comprehension task was disturbing—not what they wanted to hear. Language problems could complicate Henry's amnesia diagnosis. After my talk, Dr. Teuber requested a meeting in his office. There his tone of disapproval was unmistakable, powerful, personal and devastating. He characterized my data as "incomprehensible." For me, this was a defining moment. I wanted a PhD. I decided to pursue a thesis topic that was unrelated to Henry, putting his sentence comprehension data aside.

Dr. Teuber and I never again discussed Henry's sentence comprehension. We resumed our previously cordial relationship. Within fifteen months, I had my PhD and departed MIT at age twenty-five for a teaching career at UCLA.

Only many years later did I understand what had happened on that day in 1966. Dr. Teuber found my results "incomprehensible" because they contradicted the sacred canon, the conceptual framework that every neuroscientist takes for granted—the 1732 philosophical locomotive identified earlier as "Descartes's train." That train is what hit me in Dr. Teuber's office. Perception and comprehension were the engines powering the train. They pulled forward thinking, memory-storage, memory-retrieval, and action—the remaining cars and caboose of Descartes's train. Everyone assumed that Henry's surgery damaged only the baggage-storage car—the one that stores memories. No wonder Dr. Teuber disliked my results. My data must have suggested a train wreck, with damage extending from Henry's engine for comprehension to his cars for memory-storage, memory-retrieval, and action. Too complex! I must be wrong!

The first decades of my UCLA career were action-packed. I taught many classes. I wrote a book. I established a lab and published a line of research on how normal people understand spoken and written language—the everyday feats that challenged Henry, according to my data.

NEWS FROM MIT

In 1975, I chanced on a study that followed up on my unpublished results with Henry. The study appeared in an international journal founded and edited by my former mentor, Dr. Hans-Lukas Teuber.[1] Many of the sentences in that MIT study came from my 1966 test of Henry's ability to comprehend ambiguous sentences.[2] Its author was a fellow graduate student at MIT. Dr. Teuber was also his adviser. By authorizing that follow-up, Professor Teuber had apparently taken my "incomprehensible" results seriously.

Half of the sentences in the MIT study were ambiguous, for example, *We are confident you can make it*, and half were unambiguous, for example, *We are confident you can build it.* Henry simply indicated whether the sentence on each trial had one or two possible meanings. He could have scored 50 percent correct by ignoring the sentences, tossing a coin on each trial, and responding *one meaning* for heads and *two meanings* for tails. However, as my subsequent analyses showed, Henry scored significantly *worse* than 50 percent. He correctly responded "two meanings" to only 33.8 percent of the ambiguous sentences. Henry clearly did not understand the sentences. However, the author concluded that Henry exhibits "essentially normal" language comprehension and "the same range of linguistic capabilities" as normal individuals.[3]

The researcher had no scientific basis for calling Henry's language comprehension "essentially normal." His study lacked a control group. To correct that error, I replicated the MIT experiment at UCLA with a group of memory-normal individuals similar to Henry in age, education, IQ, and background. Those normal participants scored 81.0 percent correct—significantly better than Henry's 33.8 percent in the MIT study. Contrary to the report, Henry's sentence comprehension was not "essentially normal." It was seriously impaired, just as in my earlier ambiguity detection study.[4]

To be fair, I do not believe that Dr. Teuber or his then graduate student deliberately tried to paint a false picture of Henry's language abilities. The policy of the MIT psychology department in 1974 was to avoid statistics.

This was the unwritten rule: if statistics are needed to demonstrate significance, you should modify your experiment until you obtain an effect that is so large and obvious as to render statistics superfluous. Relative to 0 percent correct, Henry's 33.8 percent correct must have seemed large enough to count as normal.

CHANGES AT MIT

Three years after that 1974 article appeared, a tragic swimming accident in the Virgin Islands took Dr. Teuber's life at age sixty-one. Dr. Suzanne Corkin inherited his role as Henry's "keeper" and head of the MIT Clinical Research Center. It was in that capacity that Dr. Corkin deleted my lab from the list of people with access to Henry.

Here's how that happened. In 1996, my wife, Professor Deborah Burke, invited Dr. Corkin to deliver a distinguished colloquium about Henry at Pomona College. At lunch afterwards, I reminded Dr. Corkin of my unpublished 1966 results on Henry's language abilities and asked her permission to conduct three sets of follow-up experiments at MIT, with Dr. Lori James as experimenter. Dr. Corkin graciously assented and Dr. James flew to Boston six months later. All went well, and we published our initial findings in 1998.[5] The results indicated major deficits in Henry's ability to comprehend and produce sentences, and Dr. Corkin apparently learned of those deficits in February 1999, just as Lori completed the last of our tests. Lori was packing up Henry's data to return to UCLA when Dr. Corkin confronted her. Once back at UCLA, Lori forgot her exact words but vividly recalled Dr. Corkin's surprising shift in tone—from positive and friendly to cold, hard, and negative.[6] Also engrained in memory was the outcome: Lori felt unwelcome at MIT and did not wish to ever return there. Lori's three-word reply was the one touch of lightness in the conversation: "Well, there's this," she said and gave Dr. Corkin an enormous basket of California fruit and nuts—from our lab to hers. The gift did not alter Dr. Corkin's mood. She was determined to preserve Henry's status as a pure amnesic. A sinking feeling in the pit of my stomach told me that Descartes's train had struck again. I knew about the science-versus-religion battles in long-ago Italy, but assumed that late-twentieth-century MIT was different. I was wrong. Fortunately, my 1966 study plus the data that Lori collected from 1997 to 1999 gave us all the information about Henry we needed.

In subsequent years, I tried repeatedly to repair my relationship with Suzanne Corkin. When one of her graduate students informed me that Dr. Corkin was planning to visit UCLA to organize the future preservation of Henry's brain, I invited Suzanne to dinner at the UCLA Faculty Club. She declined the invitation. Her schedule was filled. I never saw her again. After reading *Permanent Present Tense* in 2013, I stopped trying to mend fences with Dr. Corkin.[7] I saw too many sentences in her book that stepped outside the bounds of science and into the realm of creative fiction—all aimed at bolstering Henry's make-believe image as a pure amnesic without language deficits.

THE LONG ARM OF THE *DSM* DEFINITION OF AMNESIA

In fairness, other labs also strove to categorize their memory patients as *pure amnesic*. The head of one California lab claimed to be studying a purer amnesic than Henry, but did not want that mine-is-purer-than-yours claim tested.[8] He failed to answer my formal request to conduct some simple tests of his patient's language abilities. When I made similar appeals to principal investigators studying amnesics in other parts of the United States, Canada, Scotland, England, and Europe, I also received non-affirmative replies. The researchers apparently considered the possibility of language deficits either too remote or too risky to warrant investigation. Poor performance on a test of sentence comprehension could land their patient in the relatively obscure diagnostic island known as *Wernicke's aphasia*. They preferred the label *amnesic*. Everyone can understand and sympathize with *memory* problems.

Still, the obsession with purity puzzled me. It was not about helping the patient. In the years following Henry's operation, there were effective clinical treatments for aphasia, but not for amnesia. Henry might have benefited from available therapies designed to improve, restore, or offset his impaired language skills. To compensate for his language problems, Henry had to invent his own strategies (see Chapter 24).

What made purity important in the *DSM* framework? Was it data on how the mind and brain work? I searched the *DSM-5* for those data without success. *DSM-5* definitions of amnesia and aphasia reflect the subjective experience of clinicians rather than laboratory tests, functional magnetic resonance imaging (fMRI) or other definitive sources of knowledge. This is the process. An elite task force of clinical psychiatrists and psychologists examines how often symptoms co-occur in an extensive population of patients. If a sufficiently large cluster of patients share the same or similar symptoms, the committee assigns a label to those patients and lists their disorder in the *DSM-5*.

One problem with this process is that *intuitions* rather than data determine what counts as a disorder and how it gets labeled. True, these are intuitions of experienced clinicians, but they nonetheless consist of hunches, gut feelings, and inner voices that say, *This is not right*—without explicit knowledge of what is right. In 1952, such intuitions inspired *DSM* experts to characterize *homosexuality* as an unnatural disorder, a label that haunted millions of Americans for decades.[9] Fortunately, intuitions change, and the current *DSM* no longer lists *homophilia* as a malady.[10]

Another problem with the *DSM* process is that symptoms characterized on the basis of intuition are difficult to communicate and apply. Studies show that the same therapist will often assign different symptoms to the same patient on different occasions, and different clinicians frequently diagnose different disorders when analyzing the same set of symptoms. Because labels such as *schizophrenic* can stain a person's reputation for life, getting a second or third opinion is advisable. The consequences of capricious *DSM* intuitions can be tragic.[11]

A third problem is that *DSM* categories such as *amnesia* lump together patients with brain damage that is diffuse, poorly understood, or unknown. In this regard, Henry was an exception. A surgical incision precisely localized in the hippocampal region of his brain caused

Henry's amnesia. When amnesia results from substance abuse, malnutrition, mercury poisoning, chemotherapy, a blow to the head, or impaired blood flow to the brain (infarctions), the brain damage is often extensive and difficult to specify. An example is *Korsakoff's amnesia* (often associated with chronic overconsumption of alcohol). Korsakoff patients typically exhibit impairment to memory, balance, coordination, and vision, with weakness, numbness, or pain in the toes, legs, and fingers, reflecting diffuse damage throughout the central and peripheral nervous systems.

Nonetheless, neuroscientists developed an effective treatment for Korsakoff's amnesia. Vitamin B-1 (thiamine) supplements or injections, together with proper nutrition and hydration, help Korsakoff amnesics learn and remember new information. Although recovery is often slow and incomplete, this stands as a spectacular achievement among existing cures for mental disorders.

Faced with the success of data-driven neuroscience contributions to understanding and treating Korsakoff's and other disorders, did psychiatrists generally abandon or at least question their intuitive approach to *DSM* categories? Not at all. A large and powerful subset of psychiatrists still prefer an intuitive rather than data-driven framework for characterizing clinical symptoms and disorders. As Dr. Daniel Barron, the resident psychiatrist at Yale University, recently noted, "They argue—absurdly—that neuroscience has little clinical relevance."[12] What framework *besides* neuroscience could possibly be more relevant to understanding brain dysfunction? Woody Allen considered the brain his second favorite organ—so what organ do those psychiatrists think they are treating: the brain, or Woody's *favorite* organ?

WHY THE WIDESPREAD APPEAL?

Given these problems, why do *DSM* definitions and diagnoses enjoy such widespread acceptance? A frequently reiterated analogy with biology explains the allure in part. Assigning the label *mammal* to an animal situates it in relation to other species and summarizes its basic biological features such as warm-blooded and feeds milk to its young. Likewise with the label *schizophrenic.* That diagnosis is supposed to situate a patient's disorder in relation to other disorders and to summarize the likely behavior of the patient: incoherent speech, inappropriate emotional reactions, withdrawal from social situations, hallucinations, expression of bizarre beliefs.

The seeming *naturalness* of intuitive descriptions adds further appeal to the *DSM* process. People often create intuitive concepts without conscious awareness. Even children spontaneously act like intuitive linguists and intuitive biologists. For example, without formal instruction, children naturally form a wide range of intuitive ideas about *words*, including ideas like *noun*, *adjective*, and *article*—long before they can explicitly describe or explain those concepts.[13] From hearing English spoken, kids know that adjectives and articles precede the nouns they modify. They say, *A beautiful day*, not *Day a beautiful*, or *Day beautiful a.*

This example highlights a third attractive feature of the *DSM* process: intuitive categories enable *predictions. DSM* diagnoses and predicts similar behaviors for patients sharing the same diagnostic label, hence *their* appeal. Likewise for linguistic intuitions. When hearing a sentence, listeners intuitively know that an adjective predicts a forthcoming noun, as does an article, a transitive verb, and a preposition. These linguistic intuitions can speed up how quickly people comprehend sentences, hence *their* appeal. The problem is that intuitions are often incorrect—whether linguistic, biological, or diagnostic.[14]

INTUITIVE CATEGORIES: A DEEPER PROBLEM

At best, all *DSM* categories are a *descriptive* summary of the symptoms that a cluster of patients tend to share. They explain neither the symptoms themselves nor why they cluster nor why some patients suffer symptoms that span several diagnostic categories. For example, the diagnosis *social communication-autism spectrum disorder* simply means that for some unknown reason, a patient shares symptoms of an autism disorder to some degree and symptoms of a social communication disorder to some degree.

However, *DSM* experts create a diagnostic category not just to *describe* a disorder, but to develop effective treatments for it, to stimulate research into what causes the disorder, and to explain how the disorder evolves over time in a patient. The problem is that the current *DSM* approach to describing symptoms and labeling disorders is unlikely to achieve those laudable goals.

To grasp this deep problem let's compare intuitive biology versus biological fact. For concreteness, let's begin with the biological intuition, common among children and untutored adults, that animals with similar physical traits prefer similar kinds of foods. This similar-bodies-similar-food-preferences idea is sometimes correct, but not always. Sharks and kingfish fit the intuition. These bodily similar species have similar diets (smaller fishes). For chimpanzees and gorillas, however, the intuition is wrong. These seemingly comparable species exhibit strikingly different food preferences. In the wild, gorillas only eat thistles, wild celery, and the shoots and leaves of specific plants and trees, whereas chimps gobble down almost any food.

Panda bears, koala bears, and brown bears illustrate another dramatic exception. Contrary to the similar-foods-for-similar-bodies intuition, these seemingly similar species prefer different foods. In the wild, koala bears only eat eucalyptus leaves, panda bears prefer bamboo leaves and shoots, and brown bears consume fish, squirrels, elk, caribou, deer, carrion, and a wide range of fruits, berries, grasses, and roots.[15]

Intuitions based on appearance inspired neither understanding nor investigation of these exceptions. Helpful insights into food preferences across species came from the decidedly non-intuitive science that developed the theory of evolution and the concepts *herbivore* and *omnivore*, based on thousands of careful studies of animal physiology and behavior related to food preferences. This is how biological scientists explained the counterintuitive exceptions. *Herbivore* species evolved digestive systems over millions of years to neutralize the poisons produced by the plants in the niche or environment in which they lived. Panda bears and koala bears are herbivores. They became picky vegetarians because the biochemistry of their stomachs evolved to handle the toxins in the specific kind of foliage that constituted their diet for millennia. By contrast, evolution gave *omnivores* general-purpose digestive systems, together with brains to help them avoid whatever foods upset their stomachs. Like humans, chimpanzees and brown bears are *omnivores*. Because their stomachs can digest virtually anything that nourishes, their habitual diet includes a variety of plants and animals.[16] By analogy, the knowledge-based theoretical approach that replaced biological intuitions with the concepts *herbivore* and *omnivore* is capable of replacing psychological intuitions with scientific concepts that allow deeper understanding and treatments for disorders such as amnesia and schizophrenia.

WHAT *ARE* YOU, HENRY?

By 1999, my research findings indicated that Henry fit a hodge-podge mix of labels. He was an impure *aphasic* (he could easily comprehend, plan, and produce familiar phrases and sentences, but not novel ones), an impure *anosognosic* (he could neither detect nor correct errors—his own or others'), an impure *anomic* (he could easily retrieve familiar words, but the rare words he once knew became gaps that he had to fill with circumlocutions and fillers such as *thing* or *stuff*), an impure *dyslexic* (he easily read high-frequency words, but made spectacular errors when reading novel sentences, nonsense-words and low-frequency words; see the next chapter), an impure *visual agnosic* (he easily recognized familiar objects in isolation, but not when they occurred in unfamiliar contexts), and an impure *amnesic* (he had difficulty learning some classes of new information, but not others; see Chapter 25).

What's wrong with this list of diagnostic labels? It is meaningless. It explains nothing. Any one of those labels ignores entire categories of other symptoms. The widely accepted DSM label *amnesia* didn't help clinicians develop a remedial plan of action for Henry and may have hindered research into Henry's condition. No solid evidence indicates that the DSM process—lumping patients into intuitive categories—advances the fundamental goals of psychology: to understand and enhance the *individual* mind and brain, in this case Henry's and other "amnesic" patients'.[17]

THE BIGGER PICTURE

The bigger picture is that Henry was a man who suffered damage to parts of the hippocampal region of his brain and was unable to quickly form internal representations for certain classes of new information. From a scientific or knowledge-based perspective, that one problem explains Henry's difficulties in comprehending and producing novel sentences, reading aloud novel sentences consisting of familiar words, reading aloud unfamiliar words and nonsense words in isolation, recognizing unfamiliar objects in isolation, recognizing familiar objects in unfamiliar contexts, detecting and correcting errors (his own and others), and recalling the unique spatial and temporal context of most types of events in everyday life.

The irony is that everyone, including Drs. Corkin, Milner, and Teuber, had a roughly accurate intuition that the hippocampus is related to memory. Their mistake was to assume that the hippocampus only creates memories for newly encountered events and facts, and *not* the memories underlying visual cognition, language and procedures or actions. True, personally experienced events are especially difficult to remember—for everyone, not just Henry. This is because the river of everyday experiences is forever new—unlike the many aspects of language memory that get repeated every day. Also true, the hippocampal region encodes novel information, but this includes any type of new internal representation, regardless of whether its content changes continually, like personally experienced events, or remains stable over time, like the way that English orders its novel combinations of adjectives and nouns. The next chapter discusses what happens in the brain when hippocampus encodes new information.

QUESTIONS FOR REFLECTION: CHAPTER 16

1) This chapter discusses problematic ways of thinking that are based on intuition or gut reactions instead of data and evidence. Can you think of similar examples from your daily life or interactions with others? How can you use awareness of this problem to enhance your own reasoning?**
2) Has a doctor ever given you an inaccurate diagnosis based on their subjective impression or bias rather than a scientific or medical assessment of your symptoms? What have you learned in this chapter that could help you correct this reasoning error?
3) Chapter 16 discusses how relying on intuition over evidence led to the mistaken classification of homosexuality as a psychiatric disorder. That episode is also an important reminder that facts can overturn a general consensus of experts. Can you think of some examples of present-day "opinions among experts" that will likely be overturned as evidence accumulates?
4) Why do you think many psychiatrists currently believe that "neuroscience has little clinical relevance"? Can you think of cases that support or refute their position?
5) Why do you think children develop stereotypes about animals? Can you identify any similar stereotypes that often persist into adulthood?
6) This chapter discusses how difficult it was to overturn the widespread belief that Henry was a pure amnesic, despite extensive evidence to the contrary. As a general phenomenon, why do you think people are so resistant to changing their beliefs?
7) What are the advantages and disadvantages of using stereotypes when trying to understand the world around you? Do you find yourself using stereotypes more or less often as you grow older?**

TEST YOUR MEMORY FOR CHAPTER 16

1) What is an intuitive linguist? Were you ever one?**
2) Why were renowned psychologists and neuroscientists reluctant to hear about Henry's language comprehension difficulties, given his well-known memory problems?
3) What is *Korsakoff's amnesia*? How can it best be treated?
4) This chapter discusses how children spontaneously learn to organize words into syntactic categories such as nouns, adjectives, articles. What is a major benefit of doing so?
5) The author discusses several parallels between how children develop concepts and how the *Diagnostic Statistical Manual for Mental Disorders* (*DSM*) develops psychiatric diagnoses. What is one of these parallels?
6) The author describes Henry, according to *DSM* criteria, as an impure amnesic, impure aphasic, impure anomic, impure dyslexic, and impure visual agnosic. What does *impure* mean in this context? What common factor underlies Henry's combination of problems?
7) Why do cognitive scientists sometimes refer to children as "intuitive linguists"?

MEMORY TEST ANSWERS: CHAPTER 16

1) p. 216
2) p. 210
3) p. 215
4) p. 216
5) pp. 216–18
6) pp. 218–19
7) p. 216

Chapter 17
THE HIPPOCAMPUS HAS A SHADOW, HENRY

WHAT MAKES YOU VULNERABLE TO ILLUSIONS?

The hippocampus does not communicate directly with the neurons that process raw sensory input, for example, the rods and cones in the retina for perceiving the visual world and the pitch detectors in the middle ear for perceiving the auditory environment. Only neurons one or more synapses above the lowest-level sensory cells convey information to the hippocampus.[1] Based on these observations, Steven Pinker, a distinguished cognitive psychologist at Harvard University, hypothesized that the hippocampus evolved to consciously process complex concepts at a non-verbal level that is higher in the mental hierarchy than the meanings of words, phrases, and sentences.[2] My research with Henry showed that the hippocampus processes letters and suffixes, a level even lower than word meanings. Nonetheless, my data confirmed Pinker's intuition about a relation between hippocampal processing and awareness: consciousness seems to arise wherever the hippocampus creates new internal representations.

THE RACE TO DISCOVER HOW LOW THE HIPPOCAMPUS GOES

When I published my first experiments with Henry in 1998, I doubted that the hippocampus only processes information at levels higher than the meaning of words and phrases.[3] My data indicated that phrase-level meanings were a problem for Henry. When asked to read ambiguous sentences aloud, Henry frequently misread words in a way that rendered the ambiguous sentences unambiguous. For example, Henry misread *John is the one to help today* as *John is the one that helped today*, a change that destroys the alternative meaning, *John is the one for us to help today*. Even when the experimenter drew Henry's attention to such errors and asked him to reread the sentence, he usually repeated his error.

Were the ambiguities in those 1998 sentences entirely responsible for Henry's reading difficulties? To find out, my UCLA lab gave Henry *unambiguous* sentences to read in four follow-up experiments. The results showed that Henry misread unambiguous sentences significantly more often than closely matched control participants.[4] Ambiguity was not his only problem.

The results also confirmed that Henry did not correct his mistakes—even ungrammatical ones such as *The boys were fed hot dogs got stomach aches*, which was how Henry misread *The boys who were fed hot dogs got stomach aches*. The missing *who* in this misread sentence was typical: for both ambiguous and unambiguous sentences, Henry often omitted short function-words like *who*, *that*, *either*, *or*, *the*, *but*, *she*, *they*, *may*, *could* and *some*—always without self-correction.

It was not that frequently used function words per se were difficult for Henry to read. He easily read them in a follow-up experiment that randomly shuffled the words in the sentences and presented them one at a time on cards. Henry's reading errors reflected a problem in integrating words into their novel context in a sentence—a difficulty related to the damage to his hippocampus rather than to his cerebellum. When a group of patients with cerebellar damage read the same sentences, they made no more reading errors than the control participants in our studies with Henry.

Explaining Henry's sentence reading errors in detail was a challenge. Traditional concepts of memory and the hippocampus did not help. His mistakes were not due to forgetting. No remembering was necessary. The sentence to be read was always there in front of Henry. An educational deficiency likewise could not explain Henry's reading errors. With the identical words presented in isolation, Henry read them correctly.

I decided to examine Henry's ability to read phrases in the unambiguous sentences that were familiar versus unfamiliar. I had my lab recruit a team of seven advanced undergraduate and graduate students conducting research on language. Their job was to assign each phrase a number between zero and five, where zero indicated *I have never encountered this phrase before* and five indicated *I have encountered this phrase very often.* Averaging the familiarity judgements across all seven raters allowed us to categorize the phrases as familiar, for example, *police station* and *hot dogs*, versus unfamiliar, for example, *cotton farmers*, and *either pork or chicken.*

Another team of researchers in my lab then re-analyzed the sentence-reading data, computing how often participants paused or produced errors within the familiar versus unfamiliar phrases. The results showed that Henry produced no more errors or unusual pauses than control participants when reading *familiar* phrases. When reading *unfamiliar* phrases, however, Henry made many more errors and paused for significantly longer than normal individuals. I concluded that the hippocampus must process novel but not familiar phrase-level meanings. Due to his hippocampal damage, Henry could not integrate the meanings of familiar words into unfamiliar phrases when reading aloud, a problem with echoes in other domains, including the recall of everyday events in their novel context of occurrence and the recognition of familiar objects in novel visual scenes (recall Chapter 13).

However, my original question remained unanswered: what is the lowest level of information that the hippocampus processes? So I proceeded to examine the next level down: Henry's ability to read unfamiliar words and nonsense words that resembled words. I asked my lab to compile three types of materials: familiar words like *trash* that children normally learn around age thirteen, long before Henry's operation; infrequently used words like *akimbo* and *abdicate* that Henry knew at one time but rarely used; and pronounceable nonsense words with no entry in the *American Heritage Dictionary.* We formed some of the nonsense words by substituting a letter in a familiar word, for example, changing *quantity* to *quintity.* Others we formed by adding an inappropriate suffix to a familiar word, for example, *friendly* became *friendlyhood.*

Like familiar phrases in sentences, Henry read familiar words in isolation without difficulty. This is because he learned the correct pronunciation for those before his 1953 operation. However, Henry misread significantly more rare words and nonsense words than did the control participants. This is because Henry could not form new internal representations for nonsense words and because the internal representations for rarely used words that Henry acquired before his surgery had become defunct.

The nature of Henry's errors was also informative. Henry often transformed rare words into familiar ones, as when he misread *remarkable* as *remark*, and he often changed nonsense words into real ones, as when he misread *ampetite* as *appetite* and *consultment* as *consultant.* Unable to form new internal representations for forgotten words and nonsense words, Henry simply retrieved existing internal representations instead.

Curiously, however, Henry often made repeated attempts to correctly pronounce both rare words and nonsense words. For example, he misread *abdicate* with a variety of speech sounds and stress patterns, first "abi•CUR•gle" (stress on the syllable CUR), then "A•bida•ckle" (stress on the initial A) and finally "abe•DI•ckle" (stress on the syllable DI). We called these *repeated approximations* because they resembled random stabs at the correct pronunciation rather than genuine attempts to correct an error. Unlike normal error-corrections, virtually none of Henry's repeated approximations ended up correcting his original error. Moreover, Henry sometimes continued his approximations after producing the correct form, apparently by accident. Unlike Henry, the control participants never produced repeated approximations and often introduced their error corrections with *repair markers* such as *I mean*, *no*, *er*, or *rather.*

My lab was pleased to discover that the hippocampal region normally forms internal representations at the level of letters and suffixes—something Henry could not do. By this time, however, other laboratories had surpassed us in the race to discover the lowest level of information that the hippocampus processes. The consensus was that the hippocampus becomes engaged when people learn new patterns of muscle movement, a level well below letters and suffixes.[5] I can illustrate this low-level procedural learning with a personal example. When I learned to play the flute as an adult, I had to master *embouchure* (pronounced "AHM•boo•shure"), an unusual way of breathing and holding the mouth that is essential for coaxing notes out of a flute. To achieve embouchure, one must inhale, then relax the central lower lip into the hole in the mouthpiece, press both lips together at the corners and sides, round the central upper lip, and blow in bursts into the hole, as if into a Coke bottle. Learning that unique configuration of mouth and lips engages the hippocampus in forming new internal representations at the muscle level and higher: the hippocampus must also link that mouth configuration with the instrument in order to prevent you from trying to blow on a trumpet using embouchure.

We likewise engage our hippocampus at high and low levels when detecting muscle-movement errors. As speakers, the high-level ideas we intend to express normally enter our awareness, but not the movements of our tongue, lips, and jaw. However, we suddenly become aware of our muscle movements when we *slur* a speech sound, perhaps due to alcohol consumption that slows neural firing and restricts blood flow to the speech muscles. If I say, "That's just falsh" instead of *That's just false*, awareness of my speech muscles turns on like a lamp. I may also become painfully aware of the implications of my slur and decide to recruit a designated driver before the party ends! The awareness that accompanies the unexpected slurry movement and the construction of that high-level action plan reflects engagement of my hippocampus.[6]

HIPPOCAMPUS AND HER SHADOW

Awareness of a slur highlights a new piece of the puzzle. Hippocampus has a shadow, a partner that follows her wherever she goes. *Consciousness* is her faithful follower. Hippocampal processing and awareness of novelty co-occur. Both come and go in tandem. Consciousness ensues when the hippocampus creates new internal representations at any level of the brain, including the muscle movement level in response to a slur. Without the creation of new internal representations by the hippocampus, however, consciousness of novelty vanishes.

Here's a common experience that illustrates the rule *no awareness without hippocampal engagement*. Imagine yourself entering a kitchen and becoming aware of the ongoing hum of an old refrigerator. Your awareness of that hum will fade and vanish from consciousness within a few minutes. This fade-and-vanish phenomenon is known as habituation. It occurs with virtually any continuous or repeatedly presented stimulus. Awareness fades once the hippocampus has formed a stable internal representation of an unchanging situation. This is because only stimuli that are unexpected or new relative to a pre-existing internal representation engage the hippocampus. Of course, this means that you will experience "absence of noise" if the refrigerator hum spontaneously ceases after a few more minutes. This altered awareness reflects the fact that your hippocampus has created a new internal representation of the changed situation.

Consciousness also vanishes at levels of processing where the hippocampus *cannot* create new internal representations. Optical illusions illustrate this phenomenon. One is the *Ponzo illusion.* In 1911, the Italian psychologist, Mario Ponzo, asked participants to judge the relative length of two horizontal lines drawn across a pair of converging vertical lines that resembled railway tracks receding into the distance. Everyone perceived the upper line as longer than the lower one. When Ponzo insisted that the lines were the same length, they were incredulous. Only after personally measuring both lines with a ruler did they disbelieve their eyes and concede that Ponzo was right. However, those measurements did not alter their perceptual awareness. The upper line still *seemed* longer. How could their hippocampus modify their belief about the length of the lines *without altering their visual perception of the lengths*? This happens because the hippocampus is directly connected to neurons in high-level cognitive systems that represent beliefs, but is not connected to neurons in the low-level sensory systems that process the cues to perspective, distance and size—the basis for perceptual awareness underlying the Ponzo illusion.

The *moon illusion* illustrates the same rule: *no awareness without hippocampal engagement*. People experience the moon as larger when on the horizon versus high overhead among the stars. Children even experience this illusion immediately after learning that the size and distance of the far-away moon remains fixed. However, this newly acquired belief does not alter children's *perception* that the moon seems to shrink as it rises in the sky. Accurate perceptual awareness of the size of the moon depends on modifying the low-level sensory processing of the cues for perceiving distance—something the hippocampus cannot do.

Sentence comprehension also illustrates the no-awareness-without-hippocampal-engagement rule. As a listener, you are normally aware of the ideas that a friend is communicating but unaware of the acoustic details conveying that message. This is because the message is generally new whereas the sound of your friend's voice is familiar. However, you immediately become aware of those acoustic aspects when a non-native speaker conveys the same message. This is because your hippocampus is hard at work in forming new internal representations of the novel auditory qualities and sounds of the speaker's unfamiliar foreign accent.

The lock-step co-occurrence of awareness and the formation of new internal representations suggests something important about how the brain works. The German-American neuroscientist Christof Koch and the British molecular biologist Francis Crick suggested that synchronized firing of neurons scattered across widely separated areas of the cortex may be the basis for awareness, perhaps entrained by neural loops from the cortex to the thalamus, a central way-station located near the hippocampus.[7] My own proposal is that awareness arises when hippocampal drivers enable two or more committed neurons in the cortex to remain strongly activated for a prolonged period (measured in seconds).[8] When I learned the name *Sarah MacPherson*, committed neurons for the familiar isolated words *Sarah* and *MacPherson* in my cortex underwent prolonged activation, a process that has two functions. One is to form new synaptic connections that link those committed neurons to an uncommitted cortical neuron that eventually serves to represent the novel conjunction of information that the committed neurons together represent (see Chapter 7). The second function of prolonged activation is awareness of that novel conjunction of information. This explains why hippocampal engagement casts a shadow. The shadow reflects the awareness function of prolonged activation, which in turn reflects the rapid, synchronized firing of neurons representing the to-be-conjoined information in different parts of the cortex, entrained by neural loops to the hippocampus and perhaps other structures such as the thalamus. The next chapter discusses the role of the hippocampus in comprehending a fundamental aspect of everyday sentences.

QUESTIONS FOR REFLECTION: CHAPTER 17

1) Which of following words, phrases, and sentences would be difficult or impossible for someone with a damaged hippocampus to process? *Hippoctopus, Hi, how are you?, Put the ice cream in the freezer, What spacetime is it?, quality not quantity, friendlyhood, appetite, consciousness, That's just falsh.*
2) This chapter discusses how the hippocampus, a crucially important brain region for understanding and remembering new information, is not directly connected to sensory processing systems (such as the visual or auditory cortices). Why do you think the hippocampus is wired in this way?
3) Recall the moon illusion. Can you think of other illusions in everyday life that are similar?
4) Recall the author's example of learning how to control the complex muscle movements for playing the flute. Can you think of some examples from your personal life? What does learning a complex skill teach us about how the mind and body works?**

TEST YOUR MEMORY FOR CHAPTER 17

1) If you suddenly lost the ability to form new internal representations, how would this affect your awareness of the world, based on what you've read in this chapter?**
2) How did research with Henry inform scientific understanding of "how low the hippocampus goes"? What has been established as

the lowest level of information processed by the hippocampus?

3) What is an example demonstrating that Henry's brain could process familiar but not novel information?
4) How is conscious awareness the faithful follower of the hippocampus?
5) Why do illusions like the Ponzo illusion or the moon illusion persist despite knowing how the magic happens?

MEMORY TEST ANSWERS FOR CHAPTER 17

1) pp. 226–27
2) p. 227
3) pp. 224–25
4) pp. 227–28
5) p. 228

Chapter 18

SHE SAID WHAT TO WHOM, HENRY?

WHY DO PEOPLE SOMETIMES THINK THEY UNDERSTAND WHEN THEY DON'T?

One day, my colleague, Dr. Lori James, asked Henry to study a Gary Larson cartoon and read its caption aloud. A bespectacled mother in the cartoon is clearly discussing her children with Edith, an older woman shown listening intently. However, Henry failed to understand that Edith was the listener. We knew this because of how he misread the punch line. Instead of reading "I tell you, Edith, it's not easy raising the dead," Henry read, "I tell Edith it's not . . . easy, the-raising the dead."[1] To omit the word *you* without correction, Henry must have thought the mother was talking, not *to* Edith, but *about* Edith, some unknown person not shown in the cartoon. Was Henry unable to grasp the most important concept in a sentence—*who did what to whom*? If so, this would indicate a serious comprehension deficit.

Henry's problem with *who did what to whom* in that caption reminded me of his special difficulties in comprehending a certain type of ambiguous sentence in my 1966 experiment back when I was a graduate student. For the ambiguous sentence *John is the one to help today*, Henry failed to recognize *John should help us* as one interpretation and *we should help John* as the other. To comprehend the two meanings in this type of ambiguity, Henry had to work out who did what to whom. He was not doing this.[2]

To find out whether I was onto something, I needed a more systematic test of how well Henry understood who-did-what-to-whom in sentences. I asked Dr. James to give Henry and a group of closely matched memory-normal controls the *who did what to whom* subtest of the Test of Language Comprehension.[3] In each trial, the participants read a sentence, then answered a multiple-choice question on the same page to assess their comprehension. A typical sentence-question pair was *The water that the mother spilled surprised the young child*, followed by *Who spilled the water: the mother, the young child, or nobody?* Henry rejected the correct answer (*the mother*) in favor of *the young child.* Overall, Henry accurately answered significantly fewer comprehension questions than the normal participants. His ability to understand who-did-what-to-whom in sentences was indeed impaired![4]

Illustration Box 18.1: UNLIKE HENRY, SCIENCE CORRECTS ITS MISTAKES

Scientists sometimes make serious errors that eventually get corrected. This is the story of how science corrected some basic mistakes mentioned earlier. As you may recall, an MIT researcher examined Henry's ability to discriminate between ambiguous versus unambiguous sentences and published his conclusion: that Henry exhibits "essentially normal" language comprehension. This take-home message was widely celebrated as confirming Henry's status as a pure amnesic, someone with memory problems but no other cognitive deficits.[5] Standard neuroscience textbooks continued to replay that message to the academic world for the next thirty years.[6]

The celebration was premature, however. The message was false. The MIT study seemed simple and elegant but contained fundamental flaws that could only be discovered by studying its procedures in detail: Henry heard a long list of sentences that were either ambiguous, for example, *The sailors liked the port in the evening*, or unambiguous, for example, *The sailors liked the wine in the evening.* Henry's task was to decide whether the sentence on any given trial had one or two possible meanings. The data showed that Henry correctly responded *two meanings* to 33.8 percent of the ambiguous sentences.

Failure to compare Henry's performance with chance was one problem with the MIT study. Henry's 33.8 percent correct was in fact significantly worse than random guessing (50 percent correct). No evidence indicated that Henry even understood the sentences.[7]

Absence of a control group was another problem. When my lab corrected that error, our results indicated significantly better ambiguity detection for normal individuals than for Henry. Far from being essentially normal, Henry's language comprehension was seriously *abnormal*![8]

Many subsequent studies confirmed that Henry's sentence comprehension was impaired.[9] Unlike Henry, science eventually corrects its basic mistakes.

The nature of Henry's *who did what to whom* errors helped explain his overall impairment. Familiar associations stored in memory dictated Henry's responses.[10] For the sentence *The water that the mother spilled surprised the young child*, Henry thought *the young child* spilled the water because existing associations in his memory linked *children* with spilling and *mothers* with mopping up. Of course, *which* specific memories tripped Henry up varied from sentence to sentence, but the effect was always the same. His comprehension of *The daughter that the mother adored fed her baby* illustrates this point. Henry said *the mother*, not *the daughter*, fed the baby in that sentence, but he was wrong again, and for the same reason as before: already formed associations linking *mothers* with feeding babies and *daughters* with being fed side-tracked his comprehension. Henry was processing the sentences using a rule that goes like this: let preformed associations in memory show who-did-what-to-whom in a sentence. Variants of this rule reappear throughout the foreign language that I call Henry-English.

To achieve genuine understanding, Henry needed to apply the usual, standard-English rule for internally representing sentences containing a main clause, e.g., *the daughter fed her baby* and a subordinate clause, e.g., *the mother adored the daughter.* This is the rule: first create a new internal representation of *who did what to whom* in the subordinate clause, then form another new internal representation of *who did what to whom* in the main clause. Applying this standard-English rule was impossible for Henry because he could not create new internal representations.[11]

Illustration Box 18.2. EVERYONE USES HENRY-RULES SOMETIMES, HENRY

You should not feel so all alone, Henry. You are not the only person who tries to comprehend standard-English using Henry-rules. We all use Henry-rules sometimes. When completing a test of general knowledge, unsuspecting undergraduates virtually always respond *two* when answering the general-knowledge question *How many animals of each kind did Moses take on the ark?* That answer is incorrect. If they had created a new and accurate internal representation of the question, they would have answered *zero* because they in fact knew that Noah, not Moses took animals on the ark. Preformed associations linking *the ark* and *animals of each kind* with *Noah* tricked them into thinking that the question was about *Noah*, not *Moses*.

Unlike Henry, however, *you* and *I* can prevent existing internal representations from hijacking our thoughts! When accurate comprehension is important, as in legal documents that must be signed, we can ensure that every word in every sentence fits together and make sense.

ONLY CONNECT, HENRY

Did Henry's difficulty with who-did-what-to-whom extend to other types of relations between the words in sentences? Were relations per se the root of Henry's problems? Was giving free reign to preformed associations Henry's only sentence comprehension problem?

To address these questions, I asked Dr. James to run Henry and a comparison group of normal individuals in a new experiment. Participants saw sentences that either did or did not contain grammatical mistakes involving time (*Yesterday Mary made it* is OK, but *Yesterday Mary make it* is not), number (*two horses* is OK, but *two horse* is not), and gender (*John hurt himself* is OK, but *John hurt herself* is not). Their task was to respond *yes* to the grammatical sentences and *no* to the ungrammatical ones.

Henry said yes to only 59 percent of the grammatical sentences, a performance level significantly worse than for the memory-normal controls, and not reliably better than the 50 percent correct he could have scored by ignoring the sentences, tossing a coin on each trial and responding *grammatical* for heads and *ungrammatical* for tails.[12]

What was Henry's problem? It wasn't relations per se. He clearly understood the relations in *What they learn about me will help others.* He reiterated those *familiar* relations for decades to anyone who would listen. Because Henry's hippocampal region damage prevented him from forming internal representations for any type of *new* information, he only had a problem with relations in *unfamiliar* phrases and sentences! The truly fatal flaw in Henry's sentence comprehension was his inability to form *new* connections. Allowing preformed associations in memory to derail his understanding was just a side effect.[13]

Illustration Box 18.3. EVEN GENIUSES CAN'T LEARN *SOME* THINGS, HENRY

What did Joseph Conrad and Henry Kissinger have in common? Both were geniuses who spoke English with a dense foreign accent. Their pronunciation obeyed a Henry-rule: use already formed internal representations to translate English consonants and vowels into articulatory movements. Joseph Conrad, the British Nobel Prize winner who many considered one of the best writers of English in the twentieth century, acquired his childhood pronunciation style in Eastern Europe. His Polish accent was so thick that his English-speaking friends could barely understand him. Henry Kissinger, the famous American Secretary of State, acquired his childhood articulatory patterns in Germany. Comedians routinely satirized his heavy German accent.

Like most people trying to learn another language after

age twenty, neither Conrad nor Kissinger ever fully succeeded in learning the articulation patterns of English. By contrast, children learning English as their first language acquire the local enunciation style within a few months. Prior to age six, children can even learn several *additional* languages without an accent. Not so, however, after age twenty, when a genetic program apparently dismantles or disengages the drivers in the brain that rapidly create the internal representations for new articulatory movements. The destruction of Henry's hippocampal drivers during experimental surgery is easily understood, but why the drivers for rapidly acquiring new articulatory information are *normally* destined to *self-destruct* or decouple at a relatively young age remains an unsolved mystery.[14]

QUESTIONS FOR REFLECTION: CHAPTER 18

1) Like Henry, we all sometimes make binding errors. An example is responding *two* to the question *How many animals of each kind did Moses take on the ark?* (See Reflection Box 18.2.) What are some other examples that you can think of in your own life? How do you think you could prevent this type of miscomprehension from happening?**
2) How do you think learning a new language as an adult might help you stay mentally sharp as you grow older?**
3) What are some similarities between Henry-English and toddler-English?
4) Why is it important to compare the test performance of a patient to chance or to a comparable control group?
6) Do you think that an undamaged hippocampus would be essential for understanding thick-accented English? Why or why not?

TEST YOUR MEMORY FOR CHAPTER 18

1) Relative to memory-normal controls, Henry had difficulty understanding who did what to whom in sentences like *The water that the mother spilled surprised the young child.* Why?
2) Why was the chance level set at 50 percent in the task of choosing possible versus impossible interpretations of ambiguous sentences? What was Henry's score on this test?
3) What key feature of Henry-English did Henry's problems in comprehending who-did-what-to-whom demonstrate?
4) What fatal flaw hampered Henry's language comprehension?
5) What were the fatal flaws underlying the claim that Henry's language comprehension was "essentially normal"?

MEMORY TEST ANSWERS FOR CHAPTER 18

1) p. 234
2) p. 237
3) p. 236
4) p. 237
5) Illustration Box 18.1, p. 235

SECTION IV
OUT OF THE BLUE CREATION

This section focuses on the blue, a mysterious region of the brain that creates new ideas with useful applications in the real world.[1]

Chapter 19
HOW MUCH HER DO YOU WANT, HENRY?

ARE YOU MORE ANALYTICAL AND LINGUISTIC OR MORE SPATIAL AND CREATIVE?

William Marslen-Wilson was an MIT graduate student in 1970 when he interviewed Henry for many hours a day over many days and created a verbatim transcript of their tape-recorded conversations.[1] His goal was to see how well Henry remembered events from before versus after his surgery. My motive in studying Marslen-Wilson's almost two-hundred-page transcript was to analyze and understand Henry's remarkable speech errors.

First, let's take a close look at how Henry spoke in the interviews. At one point, Henry mentioned that a lay teacher took over grades one and two from the nuns in the Catholic school that he attended as a boy. So Marslen-Wilson asked why she did that. This was Henry's reply:

> **Henry**: Uh . . . so that they took . . ., well . . . she . . . I say took over, and what I mean it as . . . that, as the kids progressed then they were able to . . . uh . . . they'd gone to a lay teacher . . . and they'd seen the nuns around, so when they moved to the grade, next grade, they would . . . they would naturally . . . uh . . . more eased . . . with being with the . . . uh . . . nuns than being scared . . . they were going in there as young kids, they'd be scared, right off in a way . . . but they see them around and understand them more.

What struck me about this excerpt was Henry's incoherence and lack of focus. His answers darted from topic to topic. They neither hung together nor made sense.

Would other observers share my impressions? Was Henry's incoherence normal for people of his age and background? To find out, my lab asked ten closely matched individuals with normal memory to answer three questions similar to ones that Marslen-Wilson asked Henry. My assistants then transcribed their answers onto cards, shuffled them together with Henry's answers to the same questions, and handed them to six judges—advanced students who were blind to the identity of the participants, including Henry. Their job was to score the comprehensibility, grammaticality, and on-topic focus of each answer using seven-point scales, where one meant *little*, and seven meant *much.* Overall the judges rated the answers significantly less comprehensible, less grammatical, and less focused for Henry than the control participants.

THE PROBLEM WITH FREE SPEECH

Henry's incomprehensible, ungrammatical, and unfocussed responses to Marslen-Wilson's questions showed that Henry was a not a pure amnesic. Something was wrong with his language production. However, I wanted to understand exactly how and why Henry's speech production broke down. To do this I had to determine what Henry was trying to say. For example, when Henry said, "what I mean it as . . . that, as the kids progressed . . ." did he mean to say, *What I mean is that, as the kids progressed*? If so, Henry must have substituted first *it*, then *as* for the *is* in *What I mean is* before giving up and dropping the *is* altogether. This being the case, why did Henry fail to correct those errors? Was it because he knew that Marslen-Wilson could figure out what he meant anyway? I could not answer those questions when Henry was free to say whatever he wanted.

There was a related issue. Was Henry fabricating? This was a real possibility. Henry often made up facts in his conversations with Marslen-Wilson. He even pretended to correct the facts that he fabricated. For example, when Henry asserted that someone assassinated Russian dictator Joseph Stalin, Marslen-Wilson asked where this happened. Here is Henry's reply:

> **Henry**: Three countries in a way . . . French, German, and Spanish . . . Uh . . . I said Spanish and it's not Spanish, Swedish . . .

Uh, I said X, but it's Y is a standard way to correct speech errors. However, Henry's "Swedish" is a pretend correction: nobody assassinated Joseph Stalin in Sweden or anywhere else. Henry invented both the fact and his correction.

To be fair, Henry is not alone in generating fake facts. Amnesics often describe hearsay events as personally experienced and remember things that did not happen.[2] At the end of the day, I just needed better data on Henry's use of language.

PICTURES THAT ARE WORTH THOUSANDS OF WORDS

To solve the free-speech problem, my lab switched to *constrained picture description tasks* for assessing the content and accuracy of Henry's speech. One was the Test of Language Competence.[3] On each trial of the TLC, participants see two or three words above a picture. A typical picture shows an older man looking on as a woman discusses the tracksuit a salesman is holding up. The must-use words are *although* and *wrong*. The instructions ask participants to include both prespecified words in a single grammatical sentence that accurately describes the scene. For our closely matched normal participants, this task was easy. They quickly said something like *The woman decided to buy the suit* although *it looked* wrong—a grammatical sentence that accurately characterizes the picture using both must-use words.

Henry was different. This was how he described that same sportswear scene: "Because it's *wrong* for her to be he's dressed just as this that he's dressed and the same way." Here Henry omitted the must-use word *although* and he fabricated: the only *he's* in the picture were not dressed the same, contrary to Henry's "he's dressed and the same way." Henry's ungrammatical utterance also rambled from the woman being somehow wrong to be to the attire of his unspecified *he*'s, a paragraph that violated the *just-one-sentence* rule in the TLC instructions.

Across the entire TLC, Henry omitted reliably more must-use words and produced significantly more ungrammatical sentences than the controls. Why? To find out, my lab analyzed the many uncorrected errors that rendered Henry's sentences ungrammatical. To do this, we had to determine intent—a simple task with normal speakers. Just ask, *What do you mean?* or less intrusively, note how they correct their error. If someone says, *Put the box on the desk . . . I mean, table,* they clearly intended to say, *Put the box on the table.*

With Henry, however, those royal roads to discovering intent operation were closed. Unlike normal speakers, Henry could neither correct his own errors nor state what he meant when an experimenter explicitly asked for clarification.[4] To solve this problem, my lab developed an independent set of criteria for determining the best possible way to correct anomalous utterances—Henry's or anyone else's. Illustration Box 19.1 lays out our criteria so that you can enjoy checking the accuracy of the problem solving that went into finding the Best Possible Correction for the many anomalous utterances you will encounter in the remainder of this book.

Illustration Box 19.1. Criteria for Creating the Best Possible Correction of Anomalous Utterances

Criterion 1: The Best Possible Correction is the speaker's self-correction, or failing that, their stated intent.

Criterion 2: If Criterion 1 is unworkable, propose as many reasonable ways to correct an error as possible, then score them using Procedures (a)–(d) below. The proposed correction with the highest average score is the Best Possible Correction.

Procedures for Ranking Proposals for Best Possible Correction: Assign higher scores to corrections that:

a). are more coherent, grammatical, and easily understood;
b). retain more words from what the participant actually said;
c). better fit the pictured situation and the speaker's overall utterance;
d). better fit other instances of the participant's language use.

By comparing anomalous utterances on the TLC with their Best Possible Corrections, my lab made a major discovery: Henry produced two unusual kinds of uncorrected errors called *binding errors* and *category-combination errors*. Both types of error reflect a breakdown in the uniquely human ability to combine conceptual units into words, phrases, and sentences that are new and useful. They also reveal the sophistication and elegance of the neural machinery housed in the human hippocampus.

UNCORRECTED BINDING AND CATEGORY-COMBINATION ERRORS

Binding errors occur when speakers omit essential units in a grammatical sentence. *Category-combination errors* occur when speakers combine units from categories that cannot be conjoined in a grammatical sentence. Henry's description of the following typical picture on the Test of Language Competence illustrates both types of error. A young man at a cafeteria counter is talking to the server. The young woman ahead of him in line already has her food. *Pie* and *either* are the must-use words. A typical control participant said, *I want* either *some* pie *or some cake*. Henry said, "I want some her . . . what she had." Here the words *her* and *she* could only refer to the woman at the head of the line, and the Best Possible Correction for Henry's anomalous utterance was *I want some of what she had.*

Henry's *I want some her . . . what she had* therefore included neither must-use word and contained two uncorrected binding errors, reflecting omission of the unconnected words *would* and *of* in the Best Possible Correction.

Uncorrected omissions were Henry's most common type of word

binding error on the TLC. He produced hundreds of them. They indicate that the hippocampus normally conjoins smaller concepts into larger ones in an internally represented plan that serves in this case to describe a TLC picture. Lacking a hippocampus, Henry could not conjoin the prepositional phrase *of what she had* with *some* to form the complex noun phrase *some of what she had*. Of course, one can imagine other possible explanations for unique events such as Henry's *I want some her*. Science can only explain patterns observed across many errors, and I focus here on Henry's omission of *would* and *of* in this example to illustrate the nature of hundreds of similar errors.

Henry's *some her* illustrates a *different* type of blunder. It is an uncorrected *category-combination error*. It conjoins word categories that are forbidden to marry in grammatical English sentences. *Some* is an *indefinite determiner*, a word category that can freely conjoin with *common nouns*, of which there are thousands in English. *Fun* is one of them. *Some* + *fun* is grammatical in *We had some fun*. However, indefinite determiners cannot link with *pronouns*, a category that includes *her*. *Some* + *her* in Henry's *I want some her* represents an ungrammatical word combination.

It is significant that Henry did not say *some she*, anticipating the upcoming word *she* in the Best Possible Correction *some of what* she *had*. The category-combination error *some her* clearly occurred not at the level of speech sounds, but at a higher, *conceptual level* where *she* and *her* refer to the same person. However, my lab was unable to determine whether *all* of Henry's category-combination errors involved units at the semantic rather than phonological level.

REFLECTION BOX 19.2: IS YOUR RIGHT HEMISPHERE HYPER-CREATIVE?

Many people believe that the two sides of the normal human brain usually compete, with the wilder and more creative right hemisphere kept in check by the left hemisphere—the more rational side of the brain that generates language in most people. Can individuals enhance their creativity by inhibiting their supposedly overbearing left hemisphere, thereby freeing up their super-creative right hemisphere? Neuroscience now has an easy way to investigate that issue using transcranial magnetic stimulation—a new technique that can temporarily modulate or disable a specific region or network within either hemisphere.[5]

Neuroscience also has the means to address a parallel hypothesis about Henry: that his incoherent speech reflects unsuspected damage to his left hemisphere, which in turn unleashed his visually creative right hemisphere.[6] After all, Henry's *I want some her* errs in the same way as *the bee-loud glade*, a strikingly creative phrase in "The Lake Isle of Innisfree," a poem by William Butler Yeats.[7]

Setting aside the issue of whether *I want some her* counts as creative (it does not), the idea that Henry had an especially creative right-brain is false. Henry's visual cognition was impaired in exactly the same ways as his language production (see Chapter 13).

The idea that creativity is restricted to the right-brain, the side that generates spatial images, is another myth.[8] So is the common belief that the two hemispheres of the normal brain usually compete. The truth is more complex and interesting.

In the normal, undamaged brain, *both* hemispheres *form new and useful internal representations*—the essence of *creativity*. The two hemispheres also communicate constantly with each other, and typically cooperate rather than compete. Because each hemisphere creates, stores, and retrieves fundamentally different internal representations, both can contribute to the *creative* process in the other hemisphere, thereby enhancing rather than inhibiting overall creativity.[9]

What makes interhemispheric cooperation possible is the rapid transmission of information across three information super-highways: the corpus callosum (three hundred million neurons in a mega-freeway that connects the two hemispheres), the hippocampal commissure (a pathway linking the two sides of the hippocampus), and the anterior commissure (a fiber bundle joining the left and right amygdala with the temporal lobes). Because of these anatomical expressways, signals often cross more quickly from one hemisphere to the other than from front to back within the same hemisphere.

To illustrate how interhemispheric teamwork can enhance overall creativity, recall Shakespeare's problem in communicating how Romeo felt about Juliet early in his play. His highly creative left-brain wanted to quickly establish that Romeo considered Juliet warm, bright, powerful, exciting, important, life-giving, and central to his world. However, Shakespeare knew that his audiences could not easily grasp and remember Juliet's assets in that list-like form.

To help solve this left-brain language problem, Shakespeare's cooperative right-brain almost certainly generated an image of the sun that his left-brain translated into the highly memorable metaphor, *Juliet is the sun*, which concisely summarized all of Juliet's gifts.

Both hemispheres contributed to this strikingly creative solution to Shakespeare's problem. Shutting down either hemisphere would almost certainly have curbed rather than enhanced his creativity.

Henry also produced hundreds of uncorrected *binding* and *category-combination errors* involving *suffixes* when reading sentences and isolated words aloud. For example, he misread the word *wrangler* as *wrangle* without correction, a failure to combine the verb *wrangle* with the suffix *-er*. This particular suffix-binding error is especially striking because of the contrasting meanings of *wrangler* and *wrangle*. For Henry and other speakers of American English, *wrangler* usually means "a caretaker of horses on a ranch," whereas the verb *wrangle* normally means "to have a long and complicated verbal dispute," as in *They wrangled over the details*. Together, Henry's many binding and category-combination errors involving suffixes illustrate how aging together with the inability to relearn damaged his internal representations for the form and meaning of rarely used words (see Chapter 4).

YOU ARE NOT ALONE, HENRY

Normal older adults sometimes behave a bit like Henry. In laboratory studies designed to induce errors, they produce large numbers of binding and category-combination errors involving suffixes.[10] Unlike Henry, however, older adults can and do repair their errors, and they quickly relearn, albeit not as fast as young adults. Older adults are also more coherent and comprehensible in conversation than Henry. Indeed, undergraduates rate the oral stories of older adults as more engaging than those of people their own age.

Young children often behave even more like Henry than older adults. When acquiring their native language, children often produce word- and suffix-combination errors *without self-correction*. Consider this typical word-level example spoken by a three-year-old: *The explosion died him.* What makes this category combination erroneous and ungrammatical is that *to die* is an *intransitive* verb. It cannot take an object like *him*. In contrast, *The explosion killed him* is grammatical: *to kill* is a *transitive* verb that can conjoin with objects.

Typical suffix-level examples produced by two-year-olds are *He goed* and *They wented.* These are category combination errors because irregular verbs like *go* and *went* cannot conjoin with the suffix *-ed.* Only *regular* verbs take that past tense suffix.[11] Needless to add, normal older adults never say things like *The explosion died him*, *He goed*, and *They wented.*

Despite such surface similarities, Henry and normal children produce category-combination errors for different reasons. Kids say *The explosion died him* because they assume that both *die* and *kill* belong to the transitive verb family. As soon as they acquire the transitive-intransitive distinction, they no longer make that mistake. Children likewise say *goed* and *wented* because they have yet to learn that *went* is the past tense form of the *irregular* verb *go*. When children acquire that regular-irregular distinction, they no longer say *goed* or *wented.*[12] Henry, on the other hand, says *I want some her* and *wrangle* instead of *wrangler* because aging and disuse have degraded his word- and suffix-category memories. Unable to relearn, Henry will forever persist in making binding and category-combination errors (see Chapter 4).

A WINDOW INTO THE BRAIN

Like children's verbal mistakes, Henry's unusual errors provide a window into how learning, memory, and the brain normally work. The undamaged hippocampus strongly activates two or more cortical units in specific categories in order to quickly create a new and larger cortical unit that represents their conjunction (see Chapter 10). This rule-governed combination process, also known as *chunking*, is how we humans create novel internal representations, not just for producing sentences that are new and grammatical, but for recalling novel events such as strings of unrelated words and for identifying visual objects such as a newly encountered face.[13]

Event memories combine internal representations of what happened when, where, and to whom. Face memories combine internal representations of features such as a person's nose, mouth, and eyes. Chapter 20 will continue this story, beginning with our ability to detect and correct unexpected events such as self-produced errors.

QUESTIONS FOR REFLECTION: CHAPTER 19

1) Has this chapter changed the way you think about creativity and the popular notion that the left-brain is analytical whereas the right-brain is creative?
2) Do you identify as being more analytical and linguistic or more spatial and creative? Do you think that this popular dichotomy is a useful way to categorize yourself or other people? Why or why not? Can you think of some perhaps better ways to categorize patterns of thought that avoid the myth about the two brain hemispheres competing with each other?
3) Consider the hypothetical scenario that your two brain hemispheres have aged naturally at a different rate, one hemisphere aging much more quickly than the other. Given what you've learned from this chapter (see Reflection Box 19.2), what conse-

quences would you suspect this peculiar problem might have on your ability to think, reason, use language, or remember facts and events? Would it depend on which hemisphere was aging more quickly, or would that make no difference at all?**

4) In the study that compared how Henry and closely matched, memory-normal controls answered questions about their childhood experiences, why was it important that the six judges of their language use be blind to the identity of the speakers?
5) Why do you think the myth that the left and right hemispheres compete with each other remains a popular belief? (See Reflection Box 19.2.)
6) Why do you think that evolution gave mammals two separate but interactive hemispheres, instead of just one, or perhaps three or four? (See Reflection Box 19.2.) Why do you think evolution split the brain of mammals into left versus right hemispheres, rather than front versus back hemispheres, or top versus bottom hemispheres?
7) What do you think accounts for the similar language errors and memory failures of older adults and young children?

TEST YOUR MEMORY FOR CHAPTER 19

1) What are binding and category-combination errors and how do they relate to normal aging? Regarding these types of errors, how are normal older adults like Henry and how are they different from Henry?
2) How can keeping two healthy brain hemispheres contribute to long-lasting creativity during normal aging?
3) How does Henry's freedom of speech relate to his acquired label, *pure amnesic*?
4) What two reasons made it difficult for researchers to understand what Henry was trying to say?
5) What do Henry's uncorrected word omissions, as when he said,

I want some her instead of *I want some of what she had*, indicate about the role of the hippocampus in producing language?

6) What are the three information super-highways that link the right and left hemispheres?
7) How is language production similar in normal older adults, young children and Henry? What principle saving grace do normal older adults enjoy that Henry lacks?
8) What is chunking? How does it relate to the hippocampus and creativity?

MEMORY TEST ANSWERS FOR CHAPTER 19

1) pp. 247–48
2) pp. 248–49
3) pp. 244–45
4) p. 246
5) p. 248
6) Reflection Box 19.2, pp. 249–50
7) p. 251
8) p. 252

Chapter 20
WHAT'S NEW, HENRY?

CAN YOUR HIPPOCAMPUS HELP YOU AGE GRACEFULLY?

Earlier chapters outlined some of the many ways that Henry contributed to world knowledge. For decades, researchers thought there were *at most* two kinds of memory: procedural and non-procedural. Henry showed that the brain contains not just two, but hundreds of different kinds of memory, each with its own unique form and function. Henry's binding and category combination errors indicated that word memories alone come in dozens of different varieties. Adjective memories look and behave differently from adverb memories. Verb memories look and behave differently from noun memories. When normal speakers create English sentences, their undamaged hippocampus routinely conjoins adjectives with nouns to form memorable phrases such as *beautiful day.* It also combines adverbs with verbs to create durable memories such as *happily married.* It even combines adverbs with propositions, as in this memorable sentence: *She is beautiful, happily.* However, the hippocampus *does not* routinely link adjectives with adverbs to form lasting memories. English speakers will soon forget *beautiful happily* uttered as two isolated words. There are no rules in English that would allow the hippocampus to create a meaningful phrase by conjoining the adjective *beautiful* with the adverb *happily* in that order.

Earlier chapters also reviewed how Henry taught the world what memory is. For thousands of years, people accepted Plato's idea that the brain creates faithful images of external events that resemble a photograph or the imprint of a key pressed into wax. Research with Henry showed otherwise. Memories are internal representations of events, facts, feelings, actions, concepts, and perceptions that are not like photographic images—even faded or fuzzy ones. Like most Americans who perceive and use pennies in everyday life, I cannot accurately draw one from memory. Unlike a photograph, my internal representation of a penny lacks many of the features of a real penny (see Chapter 5).

MEMORIES NEW AND OLD, FUNCTIONAL VERSUS DEFUNCT

This chapter highlights two additional concepts that Henry taught the world. One is the distinction between forming new memories versus retrieving old ones. The other is the contrast between functional versus non-functional memories. The history of my personal memory for an American penny illustrates both distinctions.

I created my internal representation of a penny as a child. When *newly formed*, that memory consisted of three features: round, copper-colored, and engraved with the head of Abraham Lincoln (see Chapter 7). Combining those three features in memory allowed me to conduct thousands of financial transactions throughout my life.

As an older professor, age sixty, I upgraded my three-feature penny memory. To conduct classroom demonstration, I learned that Lincoln's head faces rightward on a penny, the word *LIBERTY* appears behind his neck, and the words *E PLURIBUS UNUM* appear in a crest on the other side of the coin.

I retrieved those newly learned features only once—to convince my students that even *their* sharp young minds did not accurately remember them. For everyday purposes, I continued to rely on the reliable, old, three-feature memory of a penny that I had formed as a child.

Several years later, I tested my ability to draw a penny from memory. To my surprise, I retrieved the features learned as a child but not the ones added for that classroom demonstration. Those newly added features were no longer functional. Aging and infrequent use over my lifetime had rendered them defunct.

WHAT'S NEW FOR YOU, HENRY?

What's new versus old in memory was a source of confusion in early research with Henry. For the famous Canadian psychologist, Dr. Brenda Milner, any never previously encountered stimulus counted as *new*. That intuitive definition of *new* is too simple. It led Milner to falsely conclude that Henry could learn and recognize new perceptual information.[1] To see why *new* and *never previously encountered* are not synonyms, let's re-examine the incomplete figures test that Milner administered to Henry and ten memory-normal controls in the 1960s.[2]

Stimuli in that test consist of twenty incomplete drawings of familiar forms. The "fragmented elephant" in Figure 20.1 is a typical example.[3]

Figure 20.1. A "fragmented elephant" on the incomplete figures test. (Image by Micah Johnson.)

Instructions inform participants that they will see the same picture repeatedly, except with progressively less fragmentation each time. Their task is to guess what object a fragmented picture would represent if the artist had completed the drawing. At first, the pictures are so fragmented that *nobody* can guess correctly what object they represent. Eventually, however, *everybody* recognizes all of the objects. The goal is to correctly name each figure in as few trials as possible.

Milner showed that Henry and the memory-normal controls required about the same number of guesses to correctly name all twenty incomplete objects. Then an hour later, Milner asked her participants to recognize the same incomplete figures. Surprisingly, Henry's score improved significantly. Henry was learning, albeit not as much as the control participants. Their scores on the retest were significantly better than Henry's.

How did Henry identify the figures without deficit on the first test? Milner reasoned correctly that no participant taking her test had previously encountered a fragmented elephant *as an overall stimulus pattern.* Her mistake was to assume that recognizing fragmented pictures *in their entirety* was how her participants completed the task. Henry did not have to create a new internal representation of the patchy drawing of an elephant. He could correctly guess *elephant* as soon as a fragment in the progressively more complete pictures displayed a distinctive feature of an elephant, say, its characteristic trunk, tail, or tusk. After all, Henry's internal representation of an elephant that he formed as a child was still intact and functional. No wonder he could recognize incomplete figures on the first test as easily as the control participants. No new learning was required. Recognition of a familiar distinctive feature sufficed.

Why did Henry improve when retested with the same fragmented figures an hour later? The simplest explanation is that repeated use of a preformed and functional internal representation facilitates subsequent retrieval (see Chapter 9). No wonder Henry recognized the fragmented elephant in fewer trials on the retest. Repeated encounter improves elephant identification for the same reason that repeated use improves recognition of a familiar word in everyday life. Repeated neural activation gradually strengthens the existing synaptic connections between the cortical neurons required to recognize familiar objects.

Why did normal participants improve much more than Henry when Milner re-presented her fragmented figures an hour later?[4] The intact hippocampus of the normal participants formed new internal representations of the elephant fragments on the early trials of that first test, together with the way they morphed into a distinctive elephant feature and eventually into the elephant itself. By recalling this newly learned information on the second test, they could recognize the elephant in fewer trials than someone without a hippocampus for creating internal representations of the fragments and how they changed across trials. Lacking a hippocampus, Henry could not create those new internal representations of the initial fragments to facilitate elephant detection on the second test.

WHAT'S NEW CAN BECOME OLD AGAIN, AND VICE VERSA

In fairness' Dr. Milner and many other distinguished researchers in the 1960s and 1970s lacked the full corpus of Henry's responses over many tasks that was necessary to develop the current distinction between novel versus old stimuli. A *novel* stimulus requires new memory formation to carry out *the task at hand*. An *old* stimulus has an internal representation in the perceiver's brain that is *preformed* and *functional* for the current task.[5] Not just prior encounter, but *the task* determines whether a stimulus is old or new. *The person* also plays a role in keeping an old stimulus from becoming new. People have to use their existing internal representations to keep them intact and functional so that no new learning is necessary.

To illustrate how the task determines whether a stimulus is old or new, consider again my memory for a penny. For the task of accurately drawing a penny from memory, all but three of the features of my penny memory were *new* stimuli: new learning was necessary to successfully carry out the task. For the task of carrying out my everyday financial transactions, however, a penny was an *old* stimulus with a functional internal representation in my brain.

Likewise for Henry. Familiar words were *old* stimuli in the task of reading isolated words but became *new* stimuli in the sentence-reading tasks my lab gave him. The reason is that word meanings vary depending on their context in a sentence (see Chapter 8). The meaning, pronunciation, and spelling of uncommon words also represented *old* stimuli for Henry at age twenty, but after age sixty-five they became *new* stimuli over for two reasons: aging, non-recent use, and infrequent use over his lifetime degraded the existing internal representations in his brain, and lacking a hippocampus, Henry could not reconstruct his original internal representations (see Chapter 3). For Henry, those new meaning, pronunciation, and spelling stimuli could never become old and functional again.

Finally, the distinctive features of an elephant were *old* stimuli for Henry in Milner's fragmented elephant task. However, if Milner had asked Henry to accurately draw from memory an entire fragmented elephant on any given trial, it would have been a *new* stimulus, lacking an internal representation in his brain.

The lesson for science is clear. Neuroscientists make assumptions at their peril when applying common words such as *stimulus* and *new* to the brain. In the brain, what meets the eye isn't necessarily the *stimulus* that triggers behavior and *new* doesn't necessarily mean *never previously encountered*. It took decades of research in laboratories around the world to correct those mistaken assumptions and to demonstrate the critical role of the hippocampus in recognizing and learning new perceptual information (see Chapter 13).

REFLECTION BOX 20.1: HUNDREDS OF WAYS TO EXERCISE YOUR HIPPOCAMPUS

There are hundreds of ways you can exercise your hippocampus and enhance your ability to form new internal representations. One way is to practice solving new problems of the sort you are likely to encounter on a regular basis. Relish the challenge. Solving problems is good for your hippocampus. "Wait," I hear you respond in my mind's ear. "I have too many problems weighing me down already." *Lots of choices* is a selection problem that your hippocampus can *easily* solve. For *daily* workouts, your hippocampus needs one BIG PROBLEM that gives frequent birth to baby problems.

Here's an example from an older adult, ME. Failure to remember is a big problem for someone my age (seventy, plus or minus ten years). Forgetting gives my hippocampus new micro-problems to solve virtually every day. Neglecting to solve them can have serious consequences. My recent failure to remember where I left my keys is a memorable example. I could have lost most of my belongings, including my car.

After setting out on a bicycle jaunt with a friend (physical exercise *also* exercises the hippocampus), I remembered something I meant to bring on our excursion (I don't remember what that something was). I raced back to my house, leaving the front door open with my key on a keyring in the lock. My plan was to grab whatever it was, quickly lock the door, and catch up with my friend.

Finding that something took longer than expected. I exited through a door closer to my bike, forgetting about the keys in my open front door. An hour and a half into the trip, I remembered them. Visible from the street, those keys exposed most of my belongings to theft.

My first thought was to phone for help from a friendly neighbor. If she was home, she could lock my house, safe-keeping the keys for my return. The problem was, her unlisted telephone number was not on my cell phone. What to do? I hated to turn back, leaving my belongings at risk for another ninety minutes.

This problem was unlike the dozens of forgetting problems I had solved in the past. I leave Post-it cues in critical locations as prods to remember things. I note appointments in my weekly calendar. I take photos with my cell phone as a form of external memory and reminder. I consult the Web for facts I've forgotten. I know how to keep words off the tip of my tongue. To reinforce my memories, I often ask my wife if she remembers some shared experience, say, the names of people we just met at a party. Before going to bed, I once parked my morning slippers on my bathroom scale—a compelling reminder to weigh myself before trudging off to breakfast. None of these strategies could solve the new and pressing problem posed by my forgotten keys.

I decided to abort my bike outing. During my return trip, I exercised the imagination function of my hippocampus. I pictured my front door from the perspective of a thief driving past my house. Would the open door seem like a trap? The police might be inside waiting with handcuffs to arrest anyone taking advantage of this obvious opportunity.

That was an *aha* moment. Why not call the police? Just eight blocks away from my house, they could easily come and safeguard my keys and car. By then, however, I was only fifteen minutes from home. Imagining that I could probably beat the police and save myself some embarrassment, I continued to pedal.

Arriving home, I found my keys in the open door and everything as before inside the house. However, *something had changed*. My hippocampus had formed three important new internal representations:

- Never under any circumstances leave your keys in your front door. Where you park your keys is too easy to forget.
- For possible help in an emergency, become friends with your neighbors and exchange phone numbers.
- Remember that your social network can help when memory fails—- including the police.

THE HIPPOCAMPUS KNOWS WHAT'S NEW

Responding to *novelty* is a major function of the undamaged hippocampus. A wide range of recent findings shows this, including the most detailed data in neuroscience: the responses to novel stimuli of single neurons in the hippocampus, as recorded by microelectrodes, the tiny needles that neurologists sometimes implant in a patient's brain for medical reasons.[6]

Let's explore just one of those exciting new findings linking the hippocampus to novelty-reactions. Robert Knight, an MD, neuroscientist, and psychologist at the University of California, Berkeley, invited normal participants into his lab and attached electrodes to their scalp for recording *event-related potentials*, the electrical activity in the brain that external events evoke. Knight then presented two types of stimuli: old, expected stimuli versus new, never previously encountered, and unexpected ones. The event-related potentials displayed the normal brain-response to novelty: new, unexpected events triggered greater electrical activity than old, expected events.[7]

Knight next repeated the experiment with patients suffering from amnesia due to hippocampal damage. For these patients, event-related potentials were normal in amplitude for the old, expected stimuli, but significantly smaller-than-normal for the new, unexpected stimuli.[8] These lower-than-normal event-related potentials responses show that hippocampal damage doesn't just impair the ability to form new memories. It also impairs the normal brain-response to novelty.[9] Perhaps novelty-responses originating in the undamaged hippocampus trigger the process of constructing new internal representations.

ABNORMAL RESPONSES TO NOVELTY IN THE HIPPOCAMPUS

Abnormal novelty-reactions originating in the hippocampus almost certainly complicated life for Clive Wearing, the accomplished British musician, teacher, composer and Cambridge University choir master and tenor. At age forty-seven, Clive contracted a herpes simplex infection that swept through the hippocampal region of his brain, leaving him unable to remember events that happened a few minutes earlier.[10]

Unlike Henry, however, amnesic Clive experienced everything as continually and disturbingly new. If Clive's wife of many years left his side for a few minutes, Clive would greet her return with great intimacy as if she were newly arriving from far away. Even common objects were forever new for Clive. At one point, Clive's wife found him staring at a piece of chocolate. Holding it in one hand, he repeatedly covered and uncovered it with his other hand.

"Look!" he said. "It's new."

"It's the same chocolate," said his wife.

"No . . . look! It's changed. It wasn't like that before . . ."

Clive lifted and looked every few seconds, hiding and exposing the chocolate like a child practicing a magic trick.

"Look! It's different again," he said. "How do they do it?"

Even Clive's awareness of his perceptual world seemed perpetually new, as if the universe erupted into existence every few minutes. To record these cognitive explosions, Clive kept a diary of sorts, with entries such as "Proper consciousness at 2:15 p.m." Minutes later, he would cross that out and write, "NOW I am really, overwhelmingly, completely, and superlatively conscious for the first time, despite my previous claims." For decades, Clive filled page after page with these now-I'm-fully-conscious notes.

Insistent and incessant newness distressed Clive. He became agitated, distracted, and unable to sit still for extended testing.[11] By comparison, Henry was calm, passive, and undistracted at MIT, a model research participant. It was as if Henry's surgery permanently wiped out his hippocampal novelty-reactions, whereas Clive's herpes encephalitis did the opposite. It forever flipped Clive's hippocampal novelty-switch to *on.*

HIPPOCAMPAL RESPONSES TO NOVELTY: THE PERSPECTIVE FROM ERROR DETECTION

The possibility that Henry's surgery eliminated hippocampal responses to novelty that are essential for triggering the construction of new internal representations shines a clear light on the new-old distinction and on Henry's inability to detect and correct errors. To see why, let's examine the *normal* ability to detect and correct errors.

If to err is human, to self-correct is divine. Normal speakers quickly and reliably detect self-produced errors resulting in novel sequences of speech sounds in their language. By way of hypothetical example, imagine a monolingual speaker of English who intends to say, *Move your tool side-*

ways, but anticipates the upcoming /s/ in the word *sideways*, and inadvertently says, *Move your tsool sideways.* As a novel sequence with no preformed internal representation in the brain, */ts/* will trigger hippocampal novelty-responses that permit rapid error detection.

On the other hand, speakers are slow to detect self-produced speech errors that result in *familiar* sequences in their language, as when someone intends to say, *Move your tool sideways*, anticipates the upcoming */s/* in the word *sideways*, and inadvertently says, *Move your stool sideways.* As a familiar English sequence, */st/* cannot trigger a hippocampal novelty-response for detecting anything amiss. Only later, if at all, can the meaning-level contrast between the spoken word *stool* and the intended word *tool* in the speaker's sentence plan trigger hippocampal novelty-reactions for detecting the error.[12]

Nonetheless, normal speakers detect and correct self-produced errors with astonishing speed. Sometimes errors get corrected so rapidly that the word containing an error remains half-spoken, as in this example tape-recorded by the Dutch psycholinguist Pim Levelt: "the gr—. . . I mean red one." Here the speaker started to say *green one* but interrupted and corrected himself after *gr—*.[13] Two factors apparently contribute to these remarkably short self-correction times: fast novelty-reactions in the intact hippocampus and existence of the correct form in the speaker's sentence plan. Re-activating that already formed sentence plan is all it takes to correct the error once the novelty is detected.[14]

Not so for Henry, however. To say that Henry could not correct errors *rapidly* is an understatement. He could not correct errors period, whether linguistic or non-linguistic, other-produced or self-produced. When experimenters asked Henry to correct self-produced errors, he could not do so, even with unlimited time to respond. Nor could Henry detect and correct other-produced errors even when informed that something was amiss. An "impossible door" with doorknob and hinges on the same side seemed normal to Henry in the what's-wrong-here task. *She hurt himself with a knife* also seemed OK in the is-this-sentence-grammatical task. Informed that a sentence contained an error in that task, Henry still could not correct it (see Chapter 13).

The new-old distinction readily explains Henry's pattern of error-detection deficits. Other-produced errors in what's-wrong-here tasks are novel stimuli that normally trigger hippocampal novelty-responses for quickly detecting what's wrong. Not so for Henry. Without a hippocampal novelty-response, the doorknob and hinges on the same side of an impossible door are old stimuli. They activate the internal representation of a typical door that Henry formed as a child—which is why he perceives impossible doors as normal.

Self-produced speech errors are likewise unexpected and novel events for normal speakers. They trigger the hippocampal novelty-reactions for quick detection and correction. Again, not so for Henry. Without hippocampal novelty reactions, Henry has no way to perceive his self-produced errors as novel and anomalous. With no preformed plan or internal representation of the new ideas he wants to express, Henry likewise cannot correct his errors. The correct form does not exist in his brain to provide the basis for correction. How then did Henry's errors originate? The next chapter explores uninvited intruders—the basic cause of Henry's self-produced errors.

QUESTIONS FOR REFLECTION: CHAPTER 20

1) This chapter discusses how previously familiar information or objects that you used to know can become unfamiliar or novel due to aging or non-recent or infrequent use over your lifetime. How can you use this information to keep your memory sharp and obviate the need to relearn forgotten information?**
2) Why does lack of use, infrequent use, and aging degrade memories? What does this show about how the brain works?
3) Fading memories clearly create problems in everyday life. However, can you imagine why forgetting might be adaptive from an evolutionary perspective? Why are our brains built to forget?
4) What does the fascinating story about Clive Wearing, his hippocampal damage, and his permanently switched-on novelty reactions say about the relation between mind and brain?

5) Suppose you were unfortunate enough to become amnesic and had to choose between *on* versus *off* for your hippocampal novelty-reactions. Which would you choose and why?

TEST YOUR MEMORY FOR CHAPTER 20

1) What are some of the ways to exercise your hippocampus to help your memory age with grace?**
2) Why is novelty-detection such an important feature for your brain and mind and how does that feature contribute to healthy cognitive aging?
3) What does it mean for the brain to have "hundreds of different types of memory, each with their own unique form and function"?
4) How do we know that the hippocampus, though not *essential* for learning, helps speed up the process?
5) This chapter described six distinctive features of an American penny. How many of them can you recall now?**
6) Why was Henry able to perform normally on his initial encounter with the fragmented elephant test? What allowed Henry to recognize fragmented elephants in fewer trials on the retest an hour later? And why did the normal participants perform better than Henry on the retest? What do these findings indicate about the definition of *new* from the perspective of the brain?
7) What is the relation between the hippocampal novelty reactions and rapid error correction? (hint: what are errors?)
8) In what ways were Henry Molaison and Clive Wearing similar and different?
9) What brain region has been shown to be important for detecting novelty?

MEMORY TEST ANSWERS FOR CHAPTER 20

1) Reflection Box 20.1, pp. 260–61
2) pp. 259–60
3) p. 255
4) pp. 264–65
5) p. 256
6) p. 258
7) pp. 264–65
8) pp. 263–64
9) pp. 264–65

Chapter 21
UNINVITED INVADERS OF THE MIND

HOW CAN YOU PREVENT A MEMORY LAPSE FROM RECURRING?

"The unexamined mind is not," said Plato, "worth a second thought." Actually, Plato did not say that. He said (in translation), "The unexamined *life* is not worth *living*." Was this an error? Freud analyzed errors wherever he found them, including in secondhand written reports. He might surmise that the mind of someone interested in thinking must have inserted those errant words. That is true. *I* wrote those words. Freud might also suggest that my deep blue unconscious inserted those words uninvited into my awareness. It did not.

The story of how those words butted in reveals something about yourself, Henry, and normal memory. I was trying to comprehend a memory error. You may recall this incident from the previous chapter. I had forgotten my key ring in the front door of my house, briefly putting my belongings at risk, including my car. Well, I repeated that error about one week later, *again* exposing my belongings to prospective thieves passing in the street, this time for *twelve* hours. After discovering this, I felt unnerved, frustrated, fearful even. Was I losing my mind? How could I forget the seemingly *indelible* internal representation I had created after my first lapse with the keys: *Under No Circumstances Leave Your Keys In Your Front Door!* Does my unconscious harbor self-destructive Freudian impulses that intrude from the deep?

My self-doubts lasted only a minute. By analyzing that second error, what happened soon became clear. I recalled step by step my actions that evening: I used the keys to unlock my front door. After opening the door, I removed my biking gloves, closed the door, and dropped the gloves on a nearby table. Feeling a weight on my back, I removed my backpack. Mmm ... lovely curry smell from the containers of take-out Thai food inside. Did they leak during my bike trip home from the restaurant? By the time I checked, I had forgotten the key ring—still in my unlocked front door.

I next examined the basis for that memory error. I do not normally park my bicycle in my bike shed then walk around to my front door with food in my backpack. I habitually get takeout by car. No wonder I lapsed again. My old habit of entering the house through the garage without using my keys intruded and prevented me from creating the new internal representation that the situation demanded: *Unlock your front door, pocket your keys, enter the house, lock the door from inside, and ONLY THEN remove your gloves and backpack*. With that modified internal representation firmly entrenched in my brain, I felt certain I would never make that particular error again. And so far, I have not. Nor have I worried about the possibility of self-destructive Freudian impulses intruding from my unconscious. My error was simple, past tense and easy to correct. As we will see, my analyses of other errors also revealed a fundamental flaw in Freud's thinking: uncontrollable physical pressures are not what drive the brain and cause errors. I felt relieved.

REFLECTION BOX 21.1: END YOUR WORRIES, EXERCISE YOUR HIPPOCAMPUS, ANALYZE YOUR ERRORS.

Many people age forty and older worry about failures to remember.

Jeff Wheelwright, a journalist, was one of the many older adults who turn to neuroscience in search of insights into their failing memories and aging brain. He was disappointed. This is how he described his recent experiences as one of 1208 participants in a UCLA study of the aging brain.[1] He vividly remembers lying on his back in the tunnel of a cutting-edge MRI scanner, his head immobilized in something resembling a cage-fighter's mask, as the huge, multi-million dollar machine cranks up, sounding "like a cross between a jackhammer and a dentist's drill." The subsequent day and a half of cognitive and physical tests he recalled as exhausting. Then he met with the principal investigator, Professor Susan Bookheimer in the UCLA Department of Psychiatry and Biobehavioral Sciences. He wanted to understand his memory lapses, but Dr. Bookheimer never got there. She said, "You have a very nice brain." His hippocampus was "nice, fat, juicy... It's gorgeous," she added before excusing her gush. "I'm such a geek." Wheelwright asked whether this meant that his brain seemed healthier than other brains his age. She explained that "gorgeous" meant he did not move in the scanner. The images of his brain were clear and beautiful—high-quality visual reproductions. He left the lab stimulated and impressed, and then one of the research assistants asked whether he could remember how to find the garage where he parked his car.

If you are concerned about memory failure, you might help advance scientific knowledge about the brain by becoming a research participant. However, basic research is not designed to help prevent your memory lapses. Anyway, you already know some ways to do that. One is to practice analyzing your significant errors. I cannot promise that you will find this analytic skill easy to develop and use. After all, despite years of experience in analyzing errors, I failed to fully understand and remedy my *initial* lapse involving the key-ring in my front door. However, try your hand at applying what you have learned on these pages about memory errors. Analyze just one disturbing lapse soon after it happens and create a new internal representation that, with any luck, will prevent its recurrence. You'll also exercise your hippocampus and make your next major error easier to analyze and fix.

Now back to how the words *mind* and *thought* intruded into Plato's *unexamined life is not worth living*. Those words were not unconscious intruders at all. I deliberately substituted them into Plato's sentence. I wanted to create a memorable saying that might motivate you, the reader, to examine your memory errors. Plato's *unexamined* came to mind. I revised his words to "The unexamined *mind* is not worth *having*." Too negative. Too cliché. So I substituted *thought* for *having*, trying to enliven the line with rhyme and humor: "'The unexamined *mind* is not,' said Plato, 'worth *a second thought*.'" Still too negative. In *this* new context, however, *unexamined mind* contains a grain of truth: what might seem like unconscious intruders are sometimes deliberate, at least in part, and usually not deep and Freudian. As we will see, the many words that intruded into Henry's speech nicely fit this perspective.

HENRY'S INTRUDERS

The story of Henry's intruders began on the day I met him in 1966. I noticed that Henry's mind seemed to wander. Asked to describe the meaning of the sentence, *The marine captain liked his new position*, Henry's response started off-topic and strayed further off-topic: "That's why he liked the position, too, because he was above them and of all, most of all."

Where did the *most* in Henry's *most of all* come from? It feels like an uninvited intruder, an inappropriate non-sequitur triggered via free association with *of all*, his immediately prior words. The shared sounds linking *of all* to the common phrase *most of all* somehow slipped this intruder through the doors of his mind. Understanding this and many similar shallow intrusions in Henry's speech seemed important because, as we will see, normal individuals often experience similar intrusions when they produce speech errors and when a familiar word occupies the tip of their tongue.

I noticed something else right away. I felt bored listening to Henry. He endlessly repeated familiar ideas as if trapped by the words and ideas he had already expressed. On the day we met, his favorite Henry-ism was *I thought of*—which Henry repeated more than ninety times in about an hour. Other overused clichés that day were *You'd call it; I guess; I wonder; I have an argument with myself; in a way; well*...

Four years later I detected similar inflexibility of expression in Henry's conversations with William Marslen-Wilson.[2] This is Henry responding to the question, *Have you ever heard of anybody called Martin Luther King?* Note his recycled uses of the word *way* (in bold):

> **Henry**: Well, in a **way** that he . . . well . . . everything was, I guess . . . we . . . er . . . better explain it . . . the **way** . . . everything was OK for everyone else but . . . er . . . just what he's done, it's got to be just right . . . their . . . they can do anything, it doesn't make any difference, but what I do is *right*, that's . . . *it* . . .

Henry never did say whether he had heard of Martin Luther King. Instead he just recycled his *got to be right* theme—another Henry-ism that he repeated as often as *helping others.*

Confused, Marslen-Wilson asked, "So what was he saying, what was he doing?"

> **Henry**: Well, in a **way**, he was just . . . telling the people in a **way** that no matter they could think of things they wanted to and everything but . . . er . . . his **way** was *the* **way**.

Here Henry again failed to address the question and just recycled his *got to be right* theme, using *way* for a fifth and sixth time. Why did *way* and many other words intrude into Henry's speech so persistently?

THE SISTERHOOD OF UNINTENDED INTRUDERS

I know a lot about unintended intruders. When normal people in the tip-of-the-tongue state experience intruders, Professor Dan Schacter of Harvard University calls them "ugly sisters." As we'll see, however, some sisters intrude as a consequence of word-retrieval failures whereas others help retrieve the sought-for word.

What happens when you experience the tip-of-the-tongue phenomenon? You are sure you know a rarely used word, but only some of its sounds come to mind. Others seem to be playing hide-and-seek in some dark corner of your brain. You are trying to recall the word that means "an interlocking fastener made of nylon." You can see in your mind's eye how its tiny nylon hooks mesh and click together when pressed. You might hear in your mind's ear the *DAH-da* rhythm of its absent syllables—

with emphasis on the first rather than second syllable. Perhaps the sound */cr-/* enters your mind. Next may arrive the uninvited sister—*Dacron*. She *resembles* your desired word in sound and meaning, but definitely feels *wrong*. *Dacron* is a wrinkle-resistant fabric, not a fastener. You continue your search, but *Dacron* does not take no for an answer. She returns uninvited again and again. She's a pain. She torments you. It feels like she's playing a hideous game with your mind. Frustrated, you take a brief vacation from your search. Then out of the blue, your desired word softly whispers *Velcro*. Your relief is tangible. The unwanted intruder has vanished.

First observed in the 1890s by Rudolph Meringer, the Austrian linguist and collector of speech errors, unwanted sisters are now well understood. In everyday life and in laboratory studies, unwelcome sisters intrude because they are similar in sound, spelling and meaning, and identical in part of speech to your partially forgotten word: *Velcro* and *Dacron* refer to types of nylon, share the letter *O* and the consonant cluster */cr-/*, and share the same category-specific mechanism that activates all *common nouns* (see Chapter 8). Intrusive sisters delay recall of the desired word and *seem* to cause the original retrieval failure, but that's an illusion. Unwanted sisters do *not* prevent recall of the target word. They are a consequence of the tip-of-the-tongue state, not its cause. Older people suffer more tip-of-the-tongue retrieval failures, yet consistently report *fewer* unwanted sisters than young adults. If unwelcome sisters caused the tip-of-the-tongue phenomenon, more tip-of-the-tongue experiences should always mean more intrusive sisters.[3]

The actual cause of tip-of-the-tongue experiences is *memory degradation of the target word* due to aging, non-recent use, or infrequent use over one's lifetime. What helps a speaker resolve a tip-of-the-tongue experience is to encounter or produce similar-sounding words that happen to share the precise speech sounds with degraded connections in the internal representation of the target word.[4] An example is the common noun *pit* and the proper name *Pitt*. Although *dissimilar* in meaning and lexical category (common versus proper noun), *pit* and *Pitt* share all of the same speech sounds. In an experiment where participants read fill-in-the-blank sentences and later named pictures of famous people, older adults were much less likely to enter the tip-of-the-tongue state for the name *Brad Pitt* if one of the prior, fill-in-the-blank sentences was *At the center of a cherry is a p - -*. Saying the common noun *pit* unconsciously helped retrieve the proper name *Pitt*.[5]

Helpful sisters, like *pit*, only differ consistently from *unwanted* sisters in one respect: they do not share their category-specific activating mechanism with the sought-for sister. Perhaps *unwanted* sisters are trying to help just like helpful sisters.[6] The only reason an unwanted sister like *Dacron* intrudes every time you try to recall the name of her needed sister is that they both share the same common-noun activating mechanism.[7] *Dacron* is trying to set her asked-for sister free but simply lacks the right connections for restoring her degraded phonological links.

UNINVITED INTRUDERS FROM THE DEEP

Unintended intruders from the deep were Freud's specialty. An earlier chapter described a classic example that Freud read in a newspaper. A politician meant to say *battle-scarred general* but inadvertently said *bottle-scarred general*. For Freud, *bottle* was a deep intruder—thrust into the lawmaker's sentence by the unconscious and bottled-up belief that *This inept general is hitting the bottle*.

By comparison, Henry's intruders felt shallow. Many seemed to reflect nothing deeper than the sound of a word. However, I needed a better task and lots more data to fully understand Henry's intruders. I therefore turned again to *constrained picture description tasks* such as the Test of Language Competence (TLC). The results were surprising. Henry *intended* his intruders to satisfy aspects of the TLC instructions.

You may recall this typical example from Chapter 1. The TLC picture shows a man speaking to his young sons at a sidewalk intersection while pointing at a *Do not walk* sign. Henry and the control participants must produce a single grammatical sentence that accurately describes the picture and includes the words *first*, *cross*, and *before*.

A memory-normal control said, "*First* wait for the light *before* you *cross* the street" (the must-use words are in italics). Henry said, "*Before* at *first* you *cross* across." What seemed like uninvited intruders rendered his utterance incoherent, ungrammatical and difficult to comprehend. However, close inspection indicated that Henry must have welcomed those intruders. What he did was retrieve familiar words and phrases associated in memory with the target words: *at first, you cross*, and *across*, a close associate of *cross*. By tossing those old associations into a stew, Henry was able to include all three target words—as instructed. Henry was deliberately retrieving already-formed memory representations to compensate for what he could *not* do: create a novel and grammatical sentence plan that accurately described the picture—also as instructed.

Other quasi-intruders served a different purpose. This is Henry describing a TLC picture that shows a waiter in an ice cream parlor taking the orders of two female customers:

> **Henry**: Well he's putting the price of it and price of thing what it is and she wants to in there and he's waitin' to be waited on.

Here Henry's utterance is inaccurate as well as incoherent, ungrammatical, and incomprehensible: the picture shows the waiter taking orders from his customers, not waiting to be waited on. The close association with *waiter* in Henry's memory apparently intruded "waitin' to be waited on" into his speech. Similarly, "price of it and price of thing what it is" seems like a string of preformed intruders that Henry retrieved to define the concepts *it* and *thing*.

Children understand the concepts *it* and *thing*. Why would Henry want to explain those universally understood concepts to an experimenter with a PhD? In this and many other examples, Henry seemed intent on satisfying another, more basic demand characteristic of the TLC: to say something—*anything*—about the picture. This he could do—using his intact retrieval mechanisms. What he could *not* do was form a novel sentence plan for producing an accurate, coherent, and grammatical description of the picture that included the target words.

Was Freud's deep unconscious a potential breeding ground for *any* of

Henry's intruders on the TLC? Just one cherry-picked example: *I want some her*. That one example was the only reason to consider Freudian theory.

Does Henry's *some her* show that deep down, he wanted some female contact and affection? All available scientific evidence contradicts this Freudian interpretation. The Best Possible Correction for Henry's *I want some her* was *I want some of what she had*—based on the TLC picture (a man talking to the server at a cafeteria counter), the remaining words in Henry's utterance and the nature of other errors that Henry produced in that same session (see the prior chapter). Moreover, a psychiatrist diagnosed Henry as *asexual*. Henry seemed to prefer the company of men. The opposite sex did not interest him. He was emotionally unresponsive to his mother. He rejected the explicit sexual advances of a female friend in his nursing home.[8] Henry did not want that kind of her.

This is not to say that Freud's ideas cannot be tested. My former graduate student, Dr. Bernard Baars, devised a test that seemed to support a genuinely Freudian interpretation of speech errors that he induced experimentally. However, he was unable to replicate his results.[9]

STRANDED BY REPETITION BLINDNESS

Here is another possible test of the Freudian hypothesis suggested by two additional experiments conducted in my lab.[10] Both experiments examined memory under time pressure. The first showed that UCLA undergraduates report illusory intruders when recalling specially constructed sequences of words that appear briefly at the center of a computer screen. A typical sequence was WELL SMELL ART. When the computer displayed those words for one hundred milliseconds apiece (one tenth of a second), one on top of the other, the students typically reported seeing only two words: WELL and SMART. The presented words SMELL and ART vanished. In my experience they seem to go up in smoke, leaving in their place a single illusory word—SMART, an intruder that combines the SM- from SMELL with the actually presented word ART.

One aspect of this *illusory word phenomenon* is well understood: the brain cannot rapidly process repeated stimuli, in this case, the identical -ELLs in WELL and SMELL. The -ELL in SMELL disappears from the mind, a phenomenon known as *repetition blindness*. What happens next is less well understood: The brain conjoins SM-, the part of SMELL stranded by the missing -ELL (vaporized due to repetition blindness), with the final word, ART, to form the illusory conjunction SMART. Does the hippocampus create those illusory conjunctions? Nobody knows. I could have tested that idea with Henry, but Henry was no longer available for my lab to test.

ILLUSORY INTRUDERS AND EMOTION.

Now comes the experiment that would have interested Freud. It showed that normal folks are *especially* likely to recall illusory intruders that are emotionally arousing and *taboo*. UCLA undergraduates report seeing BUCK in the sequence LAKE BAKE UCK less often than seeing FUCK when LAKE FAKE UCK is presented at the same rapid rate. Both BUCK and FUCK were illusory intruders existing only in the mind of the students. However, emotions associated in the brain with FUCK made the non-word UCK easier to conjoin with the stranded F- in FAKE than with the stranded B- in BAKE.[11] If Henry and carefully matched memory-normal controls had been participants in this experiment and reported taboo illusory words equally often, I would be prepared to reconsider a Freudian interpretation of Henry's *I want some her.* However, if Henry reported taboo illusory words less often than the normal controls, his amygdala damage would make a Freudian interpretation difficult to accept because the amygdala is involved in the processing of emotional words.

REFLECTION BOX 21.2: INFORMATION DRIVES YOUR BRAIN, NOT HYDRAULIC PRESSURES

Should you worry about sexual and suicidal impulses building up in your brain and spilling over into your speech and actions? One can easily imagine how Freud's hydraulic model of mind and brain work—the idea that psychic pressure or energy builds up, bursts out, or gets diverted through alternate channels. These concepts are firmly entrenched in everyday speech. We talk of *bottled up resentment, anger welling up, anxiety surging, rage erupting, people overflowing with joy, blowing their stack, exploding under pressure, letting off steam* or *venting their feelings*. However, Freud's hydraulic theory has not panned out. Not even the hottest emotions reflect a literal buildup and discharge of *physical* energy anywhere in the brain.

Objective evidence from neuroscience indicates that internal pressures are not how your brain operates. The currency of your mind and brain is *information* not *energy*. If you feel you have made a significant Freudian slip due to uninhibited or uncontrolled psychic pressure in your brain, try to think of a more likely cause—one you might be able to counteract by analyzing and modifying the internal representation of habits related to your slip. By following this advice, you might save yourself years of expensive and time-consuming psychoanalysis.

The final idea I wanted to test with Henry concerned his amygdala—which, as you may recall, was totally destroyed in his 1953 operation. The amygdala, a brain region closely connected to the hippocampus, processes emotional information. My lab had discovered that emotional events facilitate recall of the context in which they occur. For example, UCLA undergraduates are much more likely to remember seeing the taboo word SHIT in the lower left position of a computer screen than they are to remember that the neutral word SHIP later appeared in that screen location. After replicating that basic result in numerous follow-up experiments, I concluded that in the normal brain, the amygdala sets priorities for the hippocampus: it indicates what information in memory should get bound together *first*.[12]

Did Henry's amygdala damage make it difficult for him to set priorities and make decisions? This might explain why Henry's mind seemed to wander and why he often experienced arguments with himself that never got resolved. For example, note in the following excerpt how Henry vacillated in deciding how to answer the question *Did you work in Hartford while you were still living in South Coventry?* Henry's argument with himself in this excerpt was *typical.* At sixty-six points in Marslen-Wilson's interviews, Henry expressed similar indecisiveness, always with the words *I have an argument with myself.*

> **Henry**: Well ... yes, I was ... because then ... there was another fellow then lived also in South Coventry and his father or something would pick him up as his father was coming from work ... see, there I have an argument with myself ... because just as I spa ... started to speak I thought of ... another place ... G. Fox and Company ... and ... uh ... Stately Floors, because they were two places that I'd worked at.

With Henry no longer available to test at MIT, my lab had no way of testing whether Henry's amygdala damage impaired his priority-setting emotional compass.

ONE SMALL WORD OPENS THE DOOR FOR MANY INTRUDERS

Relative to normal speakers, Henry used the tiny word *and* with abnormal frequency in describing TLC pictures. Here is Henry again trying to use the must-use words *pie* and *either* to describe the picture of the young man ordering food at the cafeteria counter: "Pie was back here *and* coffee is in there *and* this is boiled milk *and* this is not liquid." Henry's many *and-and-and* utterances were surprising. How could Henry omit the must-use conjoiner *either* but nonetheless use *and* to conjoin four independent observations in one long but grammatical sentence? Adding to the mystery, Henry's sentences disintegrated when he did use the conjoiner *either*, as in this third attempt to describe the cafeteria picture: "I want some of that pie either some pie and I'll have some." Equally puzzling was why Henry's sentences again disintegrated when he used other conjoiners such as *because*, as in this final attempt to describe the cafeteria picture:

"Coffee is in there *because* heat a solid." Coffee being somewhere *because* of heating a solid is absurd. If well-formed human thought and speech depends on the hippocampal region of the brain, what made Henry's use of *and* an exception?

The answer soon became clear. Unrelated ideas or observations conjoined with *and always* yield a grammatical sentence. Using his intact browsers Henry could retrieve *unconnected* phrases in his brain—*back here, in there, boiled milk, not liquid*—and join them with *and* into a single, well-formed sentence. However, other conjoiners such as *because* and *either* must always link *logically related* ideas in a sentence, a process that requires new memory formation—something Henry could not do.

Now I understood why Henry overused *and* relative to other ways of conjoining ideas and observations. Using *and*, Henry could satisfy the TLC instruction to produce a single grammatical sentence—albeit a run-on sentence that went on *and* on *and* on *and* on. More importantly, however, *and* allowed Henry to say *something* acceptable about a TLC picture—many acceptable somethings in fact. If Henry had attempted to create novel, accurate and grammatical sentences that included every must-use word, he would have become even more incoherent.[13]

UNCONSCIOUS BRAIN PROCESSES: PREPARING FOR ACTIVATION

I am sitting at my computer. Its two large monitors can simultaneously display different files. I wonder. My brain retrieves a word based on its meaning. Could my computer simulate that process by calling up the meaning of a word on one screen and its letters on the other?

To see whether that idea makes sense, let's say I know fifty thousand words (a modest estimate—a typical undergraduate knows about twice that number).[14] So my computer would have to store fifty thousand files representing what those words mean, plus another fifty thousand files representing their letters. The letter-file for the word *Velcro* would of course display VELCRO, but what would its meaning-file display? Not words.

Meanings are different from and more numerous than words because most common words have many meanings.[15]

Because humans can distinguish approximately ten million distinct colors, let's represent an element of meaning with a color, say, lavender for the concept *interlocking*, amber for the concept *fastener*, and cobalt for the concept *made of nylon*. My meaning-file for VELCRO would therefore display the bouquet lavender, amber, and cobalt. I can easily imagine fifty thousand computer files that display unique meaning-bouquets with links to every word I know.

Now let's program my computer to splash one of my fifty thousand meaning-bouquets onto one screen and upload the corresponding letter-file for that word onto the other screen. Is that how the brain retrieves the letters in a word? Does VELCRO just pop up when your brain activates the meaning code lavender-amber-cobalt? It's not that simple. Brains are not like computers.

The basic units in a computer are either on or off. However, neural units assume many different states reflected in their ongoing rate of firing: off, *primed* or partially activated, and intensely activated in bursts that vary from brief to prolonged. A computer hooked up to a microelectrode contacting a primed or *partially activated* neural unit emits a calm, dripping sound like rain drops falling from a roof. The same microelectrode touching an *activated* neural unit emits a staccato explosion of firing that sounds like a machine-gun (see Chapter 7).[16]

Also unlike computer units, neural units must be primed or partially activated before they can be activated for any length of time whatsoever. This preparation or *priming* stage readies the neural unit for subsequent activation—like the priming that allows an old fashioned hand pump to draw water from a well. When you prime the pump with a small amount of water, the well-water gushes out when you work the handle. *Without* priming the pump first, nothing emerges from the spout no matter how many times you pump the lever.

Why did the brain evolve this prime-first-then-activate process? Why can't neural units cut the foreplay and just gush forth with full-fledged bursts of activation? This yellow pencil I'm holding illustrates why.

When you see my pencil, your brain is primed to utter the words *pencil* and *yellow*, but you do not speak them. Only when you have occasion to express something sensible about my pencil or its color do you activate your preprimed neural units for actually saying *pencil* or *yellow*. If you immediately named everything you saw, you would qualify for psychiatric and neurological analysis. Neural priming is what distinguishes humans from automata.

Neural and computer hardware differ in another important way. Unlike computers, the brain can change itself. Neural units learn—even without help from the hippocampus. Margaret Keane, John Gabrieli, and Suzanne Corkin conducted an ingenious experiment with Henry that illustrates this unconscious learning process. Henry first read a list of familiar words for subsequent recall. When tested, Henry of course remembered none of those words. Later, in a supposedly unrelated experiment, the researchers asked Henry to identify words presented in noise over earphones he was wearing. Unbeknownst to Henry, some of the words were ones he had read (but could not remember) in the earlier memory task. The results showed that Henry recognized those recently read words much better than otherwise similar words he had not seen earlier.[17] Without his awareness, Henry's brain had learned. Saying those words left a trace somewhere in his network of unconscious memories. Repeating them will further strengthen those traces, and this progressive strengthening process explains why Henry overused familiar words and clichés in everyday speech.

Word-meaning and letter units displayed on the separate screens of my computer versus in my brain also differ in other ways. Unlike computer-files, neural units representing the meaning and letters of words interact in complex ways. The key players in these interactions are priming and activation. Both processes are unconscious. Activation that is prolonged and intense is essential for consciousness. However, the interplay between priming and activation explains why ordinary folks make errors and why uninvited sisters intrude into the mind of speakers in the tip-of-the-tongue state.

ACTIVATED UNITS PRIME THE UNITS TO WHICH THEY ARE CONNECTED

Activating the meaning unit representing the concept *made of nylon* primes or readies for activation every nylon-related word you know, including *Dacron*. Activating the letters *CR* in *Velcro* primes every word you know that contains *CR*, including *Dacron*.

To a lesser degree, *primed* (as opposed to activated) units also prime the units to which they are connected. When BATTLE gets primed, BOTTLE will also become primed, albeit less strongly than if BATTLE had been activated. The diminishing magnitude of second-order priming is essential because of how neural units get activated. Browsers always activate the most primed unit in their category, fiefdom, or domain—in this example, the set of common nouns that a speaker knows, including BOTTLE. Because of this most-primed-wins activation procedure, intrusion errors are rare. A sentence plan that includes the concept *battle* will strongly prime BATTLE, so that the common-noun browser will normally activate BATTLE, not BOTTLE, as the most primed neural unit in its domain.

However, priming can come from anywhere, not just from an activated sentence plan. Seeing a bottle can prime BOTTLE. Ongoing internal speech can accidently prime BOTTLE—especially when discussing or thinking about an alcoholic who is prone to *hit the bottle*. This *extraneous priming*, accumulated from elsewhere, might occasionally make BOTTLE more primed than BATTLE, so that the next thrust of the common-noun *pump-handle* forces the speaker to say *bottle-scarred general* instead of *battle-scarred general*. The idea that some unconscious stuff can accumulate from elsewhere was central to Freud's thinking[18]—a terrific insight in 1901—but neither an aggressive impulse nor a pent-up urge to "tell it like it is" is needed to explain Freudian slips.

The next three chapters continue this story, showing how the most-primed-wins activation principle and its companion, the organization of neural units into categories such as common versus proper nouns, explain the creative power of memory and language.

QUESTIONS FOR REFLECTION: CHAPTER 21

1) Dr. MacKay discussed how he came to leave his keys in his front door. Have you ever experienced a similar problem? Can you recall the events leading up to that mishap? Can you think of some ways to ensure that mishap doesn't happen again?**
2) Are you concerned about memory failure? Practice analyzing your lapses to understand what went wrong. Do your analyses help prevent similar memory failures or give you some ideas for how to possibly prevent them next time? (See Reflection Box 21.1.)**
3) This chapter explains how normal folks are especially likely to recall illusory intruders that are emotionally arousing or taboo. It also describes how emotional information interacts with memory to facilitate future recall. Can you think of ways to use this information to enhance your own memory?**
4) Taking cognitive tests and getting an MRI of your brain can sometimes provide insights into your behavior. Reflection Box 21.1 discussed journalist Jeff Wheelwright's participation in a cognitive neuroscience study. Have you ever participated in a similar study? If so, what were your reasons for doing so? If not, why might you want to?
5) What do these statements mean to you: "The unexamined life is not worth living." (Plato); and "The unexamined mind is not worth a second thought." (Dr. MacKay)?
6) What does neural priming tell us about how unconscious thoughts can influence our conscious mind?

TEST YOUR MEMORY FOR CHAPTER 21

1) What are uninvited intruders and how can they sometimes actually help to remember temporarily forgotten words?
2) How is a brain *not* like a computer and how might these differences help as you grow older?

3) If you experience a memory lapse, what can do to avoid making the same lapse in the future?**
4) What is the primary cause of the language and memory error known as the tip of the tongue phenomenon?
5) What are unwanted sisters and why do they intrude when people are in the tip of the tongue state?
6) What evidence indicates that unwanted sisters are a symptom and not the cause of tip of the tongue experiences? (hint: differences between younger and older adults)
7) What is the illusory word phenomenon, when does it occur, and what does it say about how the brain works?
8) Why did Henry overuse the word *and* but not the word *because*?
9) What are the prime-first-then-activate and the most-primed-wins principles and how do they work in the brain?
10) In what ways does neural hardware differ from computer hardware? What do these differences indicate about how humans versus computers process information?

MEMORY-TEST ANSWERS FOR CHAPTER 21

1) pp. 273–75
2) pp. 282–83
3) Reflection Box 21.1, pp. 269–70
4) p. 274
5) pp. 273–74
6) p. 274
7) p. 277
8) p. 280
9) pp. 282–84
10) pp. 282–83

Chapter 22
HOW CREATIVE ARE YOU, HENRY?

ARE YOU VERBALLY CREATIVE?

"Laurence Welk is a wonderful man. He used to be, or was, or—wherever he is now, bless him."

"Definitions, heck with it . . ."

"Our enemies . . . never stop thinking about new ways to harm our country and our people, and neither do we."

"There's an old saying in Tennessee—I know it's in Texas, probably in Tennessee that says, 'Fool me once, shame on . . . shame on you. Fool me . . . You can't get fooled again!'"

Does that fractured syntax remind you of Henry? Those incoherent, incomplete, inaccurate, and ungrammatical sentences are public pronouncements of George W. Bush, former president of the United States. Some people thought the president was not playing with a full verbal deck.[1] Others believed he cooked up his linguistic stews in order to seem folksy, fallible, and endearing—someone who millions of ordinary Americans could relate to and vote for. If so, this would qualify his incoherent and ungrammatical statements as creative.

Were Henry's broken utterances also creative? The book about Henry by Dr. Suzanne Corkin makes eleven separate anecdotal references to Henry's keen sense of verbal humor.[2] If creativity is the soul of wit, then Henry might have been creative in everyday life—unlike in the well-controlled conditions of my laboratory studies. After all, Henry's novel utterance *I want some her* was only *inappropriate* because the TLC task demanded a *grammatical* sentence. In ordinary conversations, novel sentences that are ungrammatical, unusual, and incomplete might serve useful functions such as brevity, authenticity, and ease of communication. They would only count as non-*creative* if they disrupted real-life communication.

HENRY'S VERBAL CREATIVITY

Was Henry as creative as normal speakers in real-world interviews? To find out, I obtained the transcripts of face-to-face interviews with sixteen minor celebrities plus Henry, who was also a minor celebrity, having starred on a National Public Radio program. The normal interviewees included Michael Jackson, John Cage, and Maya Angelou, with Pamela Grundy, Oprah Winfrey, and David Gregory among others as interviewers.[3] Henry's interviewer was Dr. William Marslen-Wilson, a graduate student at the time.[4]

Although all of the interviews were biographical, they differed in significant ways. Interviewers of the normal celebrities asked easy and difficult questions about equally often. *Difficult* questions requested a considered opinion, for instance, *What does that say about West Charlotte, do you think? Easy* questions invited yes/no answers about a simple fact, for example, *Have you been in Canada before?* William Marslen-Wilson asked Henry thirty-five times as many easy questions as difficult ones. He was clearly trying to help Henry out.

Henry and the normal celebrities also asked questions. Some were content questions requesting additional information from the interviewer, for instance, *Which time are you referring to?* Others were comprehension questions calling for help in understanding the interviewer's question, for example, *What do you mean?* Normal interviewees mainly asked content questions. Henry mainly asked comprehension questions. He was clearly struggling to understand—the antithesis of creative comprehension.

Henry's responses to Marslen-Wilson's questions also showed no sign of creativity. His remarks were the opposite of creative. They disrupted rather than facilitated ongoing communication. For example, unlike the normal interviewees, Henry usually veered off topic when his interviewer made a request. In this excerpt, Marslen-Wilson asked Henry to explain why he considered Martin-Luther King "a seditionist." Note how Henry shifted the topic from sedition to Richard Nixon, the president of the United States at the time:

Marslen-Wilson: How do you mean [Martin-Luther King was "a seditionist"]?

Henry: Well, in a way that he ... well ... everything was, I guess ... we ... er ... better explain it ... the way ... everything was OK for everyone else but ... er ... just what he's done, it's got to be just right ... their ... they can do anything, it doesn't make any difference, but what I do is **right**, that's ... it ... [emphasis in the original]

Marslen-Wilson: I'm not ... so what was he saying, what was he doing?

Henry: Well, in a way, he was just ... telling the people in a way that no matter they could think of things they wanted to and everything but ... er ... his way was **the** way. [emphasis in the original]

Marslen-Wilson: What was his name?

Henry: I think of Nixon right off.

Henry also disrupted communication by driving conversations into circles, as in this later excerpt. Note how Henry claimed to love classical music, then circled the topic away from, and then back to that love in response to Marslen-Wilson's calls for clarification.

Henry: Well ... believe it or not ... I liked the long-hair music.

Marslen-Wilson: Oh yes?

Henry: Uh-huh ...

Marslen-Wilson: Which kind of long-hair music?

Henry: Well ... in a way, I used to play the records too that we got ... of the ... well ... the ... orchestras they have, like in Philadelphia ... and ... uh ... the big orchestras ... that way ... the ... the

. . . long-hair stuff . . . with the tens and the twelve-inch record . . . [chuckles]

Marslen-Wilson: Any . . . any particular composers . . . or any particular type of long-hair music?

Henry: Well, just as you said of composer I thought Kern right off . . .

Marslen-Wilson: Who?

Henry: Kern.

Marslen-Wilson: How do you spell that?

Henry: K-E-R-N.

Marslen-Wilson: Oh yes, I know . . . Jerome Kern? [a composer of *popular* rather than classical music]

Henry: . . . And that's what I was thinking of, I was trying to think of the first name and I couldn't.

Marslen-Wilson: That's the guy you mean? . . . is it?

Henry: . . . It is and it isn't . . .

Marslen-Wilson: It might have been somebody else?

Henry: It might have been somebody else . . . because I think of . . . uh . . . Jerome Kern as . . . uh . . . playing in an orchestra. [In fact, *Jerome Kern* did not play in orchestras.]

Marslen-Wilson: What sort of music is this? . . . I mean . . . operas . . . symphonies . . . musicals . . . what?

Henry: I guess you could go the whole length, all of them . . . I wasn't very . . . I wasn't particular . . . about . . . well there was . . . uh . . . I guess you could say the mid-twenties that . . . uh . . . that kind of music I didn't care for at all. [DM: *Note Henry's shift here away from the original topic—his alleged love of classical music.*]

Marslen-Wilson: Jazz?

Henry: Jazz . . . that kind . . . I like the . . . uh . . . in a way . . . the symphony music.

[DM: *Note Henry's return here to the original topic: his love of classical music.*]

By disrupting his real-world conversations, Henry's circular spirals were clearly uncreative. They nonetheless demonstrated a remarkable ability to remember conversational themes. After nearly twenty back-and-forth speaker-turns that addressed other issues, Henry recalled the original topic of conversation: his love of classical music. Henry did not produce off-topic comments because he forgot the current topic of conversation.

All of Henry's contributions in the 178 page transcript of conversations with Marslen-Wilson qualified as non-creative. This finding raised a question. Was Suzanne Corkin right about Henry's "keen sense of humor"?[5] Could Henry really create and comprehend original wit and humor in everyday life? The next chapter addresses that issue, together with the question of whether humor can enhance the mind and memory of normal older adults.

QUESTIONS FOR REFLECTION: CHAPTER 22

1) To what degree do you think that Henry was aware of his speech and memory problems? Do you think that increasing his awareness would reduce his problems? Explain why or why not.
2) Have you ever noticed fluctuations in your verbal creativity over the course of a day or at different stages of your life? What might contribute to these changes?
3) Do you think that verbal creativity implies greater artistic, musical, or athletic creativity? Why or why not?

TEST YOUR MEMORY FOR CHAPTER 22

1) What are some useful functions that incomplete or ungrammatical sentences might serve in ordinary conversations? What would make them either creative or noncreative?
2) In what ways was Henry's speech and that of former president George W. Bush similar?
3) When being interviewed, Henry more often asked comprehension rather than content questions. What is the significance of that finding?

4) What is the significance of Henry's remarkable ability to remember the topic of conversation in interviews?

MEMORY-TEST ANSWERS FOR CHAPTER 22

1) p. 288
2) p. 287
3) p. 288
4) pp. 288–89

Chapter 23
YOU'RE NOT KIDDING, HENRY

BEWARE. LAUGHING CAN BE A WARNING SIGN OF HEALTHY AGING.

I love humor. I love the brevity. I love the ambiguity. I love the echoing analogies in this pithy poem by Richard Armour.

> Shake and shake
> The catsup bottle.
> None'll come—
> And then a lot'll.[1]

This is not just a light-hearted meditation on catsup. Armour is also invoking deeper themes about persistence and the many things in life where none'll come and then a lot'll. Affection. Love. Sex. Knowledge. Friends.

Do you have a keen sense of humor? If so, count yourself lucky. Your emotional brain and hippocampal region are intact, and you can form new and useful internal representations in your cortex. You can also comprehend and perhaps even create jokes like these:

Husband*:* I don't think our son is absent-minded.
Wife*:* Have you ever *met* our son?

Larry: Seven days without a pun makes one weak.
Liz: Why is punning so rewarding?
Larry: A pun is its own re-word.

Wife: Why is divorce so expensive?
Husband: It's worth it.

Mark Twain: There's one way to find out if a man is honest: ask him; if he says yes, then he's lying.

Woody Allen: My one regret in life is that I am not someone else.

HENRY'S "KEEN SENSE OF HUMOR"

According to some researchers, Henry understood jokes like these and generated similar one-liners of his own.[2] But how could that be? Creating new and *unfunny* statements was beyond Henry's capacity.[3] Was humor somehow an exception? Did Henry really have what it takes to create and comprehend original wit and humor? Understanding ambiguous sentences was difficult for Henry. So how could he comprehend *Seven days without a pun makes one weak* and the many other jokes with ambiguity at their core? Henry also did not understand metaphors. So how could he catch the metaphoric echoes in Richard Armour's humor? And what about the amygdala damage that stunted Henry's experience of emotion? Surely humor engages emotion.[4]

Again something was amiss and big issues were at stake: not just the neural basis of creativity and humor, but the validity of my experiments on Henry's ability to comprehend and produce sentences. I needed to determine whether the stories about Henry's "keen sense of humor" were true or whether jokes contradicted everything I knew about Henry.

DID YOU FAKE IT, HENRY?

Searching for a solution to this puzzle, I discovered an unpublished study on Henry's sense of humor. In 1970, researcher William Marslen-Wilson asked Henry to explain why a particular captioned cartoon was funny.

In the cartoon, a distraught mother is trying to escape the chaos in her kitchen; unwashed dishes in the sink; soapy water overflowing onto the floor; toys and dirty clothes strewn about; four young children squabbling and crying. It seems like she might be dying of thirst in the desert, gasping "Water! Water!" but the caption reads, "The Pill! The Pill!" The troubled woman clearly craves contraception.[5] She has too many children.

Henry's amnesia predated the invention of contraceptive pills by many years. Lacking an internal representation for this key concept, Henry had no way to understand the joke. He could not connect "the Pill" with the mother's craving. For Henry, this was the joke from hell. It's no wonder he did not get it.

What's astounding is that Henry pretended he did get it. This was Henry's response when Marslen-Wilson asked him to explain what was funny:

> **Henry**: Well . . . it's a wonder of the . . . uh . . . the mother of course going out of the room . . . but seeing "The Pill! The Pill!" and all the . . . like soap suds in a way that there's been raised there . . . she can't do ANYthing, however, she has to do EVERYthing . . . she . . . both ways of looking at it . . . [emphasis in the original]

Puzzled, Marslen-Wilson asked, "Who is saying . . . 'The Pill! The Pill!'?"

> **Henry**: It's the little girl that's saying to the boy . . .

Henry clearly did not know who said the caption, but Marslen-Wilson played along and asked why the girl said that. This was Henry's cooked-up response:

> **Henry**: Well . . . to . . . point out to the boy that that's what it was that . . . the little pill that the mother possibly had dropped in to make the soapsuds and . . . and maybe . . . she thought maybe, well, it was more than one pill that she had put in, and that got . . . that's why she'd got so many.[6]

Instead of concocting this story, Henry could have asked Marslen-Wilson what kind of pill the mother wanted, but swimming in a sea of non-comprehension, Henry may not have realized that he did not understand. Humor was the least of Henry's problems.

WHAT'S FUNNY HERE, HENRY?

If Henry pretended to understand an incomprehensible cartoon in the laboratory, perhaps he also faked a love of humor in real life. The discrepancy between my experimental data and the anecdotes about Henry's "keen sense of humor" seemed worth pursuing.

This was the key question: can Henry comprehend humorous cartoons that he could *in principle* understand? To find out, my postdoctoral fellow and I collected a set of captioned cartoons containing words that Henry knew before his surgery, and on her next trip to the Center for Clinical Research at MIT, Dr. James handed Henry the cartoons across a table and asked him to explain why each was funny. She brought back Henry's tape-recorded accounts for analysis at UCLA.[7]

We discussed *The Far Side* cartoon, "Raising the Dead," by Gary Larson earlier. It depicts two women chatting in the living room—a common scene except that both are *ghosts*. Edith is listening as the other ghost woman complains about her mischievous ghost children, Billy and Sally. Billy is shown flitting aimlessly around the living room; Sally is floating head first down some stairs.

The caption reads: "Oh, I don't know. Billy's been having trouble in school and Sally's always having some sort of crisis. I tell you, Edith, it's not easy raising the dead."

Henry did not find that situation funny. Trying to help, Dr. James asked him to read the caption aloud. Henry read: "Oh. I don't know. Gus (pronounced *Guzz*) having—been having trouble in school and Sally having sh—always having some sort of crisis I tell Edith, it's not . . . easy, the—raising the dead."

Surprised, Dr. James asked who *Gus* was. "This is a—an older man," explained Henry, pointing at the picture, "and Billy is the name for a very young kid." Henry apparently mistook Billy, the boy ghost swooping around the room and having trouble in school, for an older man who in his opinion should be called *Gus*.

We do not know why Henry thought an older man might be attending school and having trouble. However, Henry's larger error—uncorrected omission of the *you* in *I tell you, Edith*—points to a conceptual train wreck. Henry clearly thought the ghost mother talking *to* Edith *was* talking *about* Edith—some unknown person or ghost not shown in the cartoon. Unaware of who was talking to whom, it is no wonder Henry could not link Billy and Sally's problems to their mother's difficulty in raising the dead (children)—the core of the joke.

There's more. The gist of our cartoons escaped Henry. He even failed to understand that the ghost cartoon was about ghosts. Noting that he could see through one ghost woman to the armchair she was sitting on, Henry suggested that the cartoonist had drawn her wrong. He complained that she (the cartoonist) just "bl . . . the . . . blackens the whole way, and everything . . ." using some kind of "blackening rule." For Henry, a new kind of Picasso might have drawn the cartoon.

Henry's ability to comprehend the cartoons was seriously impaired, a finding that directly challenged the anecdotes about Henry's keen sense of humor. Unlike Henry, the participants in our normal control group always understood what the cartoons were about, never fabricated facts resembling Henry's blackening rule, and invariably highlighted some genuinely funny aspect of the cartoons. All found it amusing that ghost mothers might experience difficulties raising ghost children—just like real mothers raising real children.

SAY SOMETHING FUNNY, HENRY!

Given Henry's comprehension deficit, it seemed unlikely that he could generate humor on his own. Inability to comprehend humor would be a roadblock to creating humor from the outset.

My lab nonetheless looked for signs of humor in the 178 pages of transcribed conversations between Henry and William Marslen-Wilson.[8]

The search was futile. Henry made not one witty or amusing comment. On the contrary, his remarks were the antithesis of wit—incoherent, halting, vague, redundant, rambling, boring, and incomprehensible. Henry clearly did not have what it takes to deliberately amuse others: clarity, aptness, brevity, precision, a sense of timing, and the ability to make every word count.[9]

Henry also lacked the neural basis for generating wit. Consider the one-liner that began this chapter. In response to her husband's denial that their son was absent-minded, the wife quipped, "Have you ever *met* our son?" This retort reflected the formation of three new internal representations, roughly: *Despite knowing him since birth, my husband thinks our son is not absent-minded*; *Everyone else who meets our son thinks he's absent-minded*; and *To suggest that my husband has never met our son would be amusing.*

Those novel internal representations were useful and therefore creative. Her teasing comment commanded respect for her intellect, it challenged her husband's blindness to their son's absent-mindedness, and it almost certainly forced him to re-evaluate his son's behavior.

Comprehending humor likewise requires creativity. To get his wife's joke, the husband had to form three new internal representations, roughly: *After meeting my son, other people believe he's absent-minded*; *My wife knows I've met my son but is suggesting otherwise*; and *That's funny.*

Lacking a hippocampus, Henry could create *no* new internal representations, let alone creative ones for composing and comprehending quips such as *Have you ever* met *our son?*

REFLECTION BOX 23.1: RELISH THE REWARDS OF HUMOR AND LAUGHTER

With the hippocampal region as its basic driver, humor should enhance memory, and it does.[10] Research indicates a link between humor and slower rates of memory decline in older adults, perhaps because laughing inhibits the release of the stress hormone *cortisol* into the bloodstream, thereby limiting the damaging effects of stress on the hippocampus. Laughing also reaps emotional rewards. Humor enhances positive moods and feelings of well-being. Shared with friends, family, and co-workers, humor can strengthen group bonds.

Humor also benefits health directly. It reduces chronic pain associated with arthritis, perhaps by releasing *endorphins* into the blood stream (the natural painkillers produced in your body). Like exercise, humor also lowers blood pressure, reduces muscle tension, and stimulates circulation, perhaps because the muscle convulsions of laughter exercise the lungs, abdomen, face, and upper body. Laughing even improves sleep, perhaps by facilitating digestive processes, boosting the immune system, and reducing allergic reactions.[11]

So do what Henry could not do. Create a diary or playbook of the funny events in your life. Seek out opportunities to generate and enjoy humor. Share amusing anecdotes with your friends. Form a quip club with prizes for the funniest joke in prespecified categories. If your health care provider offers a laughter therapy group, join it. Turn life into a fun-fest. Unleash your lighthearted self. Engage in serious nonsense. Enjoy the many benefits of glee.

DISSECTING FROGS: ANECDOTES UNDER THE MICROSCOPE

Nothing more sharply distinguishes scientists from news reporters than their attitude toward anecdotal evidence. For scientists, personal reports from everyday life represent the lowest form of evidence. Researchers prefer one well-controlled experiment to any number of anecdotal accounts. Galileo wanted better evidence for the shape of the earth no matter how many worthy people swore that the world out their window was flat. For scientists, understanding what's going on is more important than surface appearances, engaging stories, and sincere opinions.

Science reporters and journalists, on the other hand, prefer a vivid and memorable tale from everyday life to any number of well controlled but difficult to comprehend observational analyses. The writers Mark Twain and E. B. White especially deplored attempts to analyze jokes. They considered close scrutiny of humor as on a par with dissecting a frog: the wit dies in the process and leaves an ugly mess on the table. They also insisted that one must be funny when writing about humor. For a scientist, however, this is like demanding that a researcher have cancer in order to study its behavior.

Leaving unanalyzed the anecdotes about Henry's wit was not an option for me. If a "keen sense of humor" was the only logical explanation for any one of them, then people could reject as artificial all of my experimental observations to the contrary. I needed to show why personal stories about Henry's sharp wit represented inadequate evidence.

I began by putting the most recent anecdote about Henry's sense of humor under the microscope, working backwards to the oldest in the literature. My dissections indicated that all of the stories had better explanations than a "keen sense of humor." My analyses also confirmed the importance of rock-bottom scientific procedures: creating audio recordings to rule out mishearing, asking questions to establish what participants understand, considering alternative hypotheses.

Most of the anecdotes focused on Henry's keen sense of *verbal* humor. This surprised me given Henry's well-established deficits in understanding ambiguous and metaphoric expressions, common elements in verbal humor. But the verbal versus non-verbal distinction did not matter. None of the stories held up.

THE "HI JOHN" STORY

In the "Hi John" story, one of Suzanne Corkin's lab assistants engaged Henry in a prank. Henry was to say, "Oh. Hi, John" to whomever entered the lab next. After helping Henry practice this line for a few minutes, the assistant left the room and quickly returned with John, a graduate student who knew that Henry was profoundly amnesic and did not expect a prank. John's amazed expression when Henry recalled his name gave Henry and the assistant a good laugh.[12] But did Henry actually get the joke? Laughter without comprehension is both common and contagious.[13] Compliant as usual, Henry may have laughed along with the assistant without understanding why John's amazement was funny. The creator of the prank clearly had "a keen sense of humor" but her anecdote does not prove that Henry did. She did not ask Henry why he was laughing.

THE PUZZLE KING STORY

The Puzzle King Story was the most frequently repeated anecdote to come under my microscope. One day, Dr. Corkin encountered Henry working on a crossword puzzle and commented: "Henry, you're the puzzle king."

"Yes," said Henry, "I'm puzzling."[14]

Was Henry's reply a creative quip? Did he really understand Corkin to mean *You rule, Henry. You're a king among puzzle solvers*? Was Henry trying to advance the conversation with a double reference to his crossword puzzle habit and his existential condition—the profound amnesia that puzzled him for most of his life?

This clever quip idea is exactly the sort of hypothesis that scientists shun. It contradicts the fact that metaphors resembling Corkin's *puzzle king* were especially difficult for Henry to comprehend (see Chapter 15). It also presumes complex, high-level processes that are poorly understood and lack independent support. To verify her clever quip hypothesis, Corkin needed to ask Henry what *puzzle king* means. She did not.

Under an alternate account of the puzzle king story, Henry inadvertently collapsed the words *puzzle* and *king* into *puzzling*. Focused on his puzzle, he simply misheard Corkin's *You're the puzzle king* as *You're, uh, puzzling*. Under this hypothesis, Henry's *Yes, I'm puzzling* is a normal, appropriate and unfunny response to a misheard statement.

Why is this alternate narrative preferable to Corkin's clever quip idea? Mundane mishearing is exactly the sort of low-level hypothesis that science prefers. Moreover, that uninteresting hypothesis enjoys independent support. Careful analyses of tape-recorded conversations indicate that Henry and other amnesics often confuse *the* and *a*.[15]

THE LONG DRIVE STORY

A lengthy automobile trip from New Haven to MIT was the setting for the long drive story. The driver was Professor Hans-Lukas Teuber. His passenger was Henry.[16]

"That was a long drive," commented Teuber, "Are you stiff?"

"Nope," replied Henry, "I haven't had a drop."

Did Henry comprehend and then deliberately ignore Teuber's obvious reference to physical stiffness? Did he intend to lighten the mood by raising the topic of being intoxicated (a now obsolete meaning of *stiff* seen in the slang expression *Dat bum's a stiff*)? Was Henry's remark a creative quip?

Henry's comment fails to qualify as either creative or quip-like. Because of his amygdala damage, Henry had difficulty perceiving his own psycho-physiological states and emotions[17]—a condition that rendered him unable to feel his moods or to evaluate his state of physical stiffness. Henry had no way of responding appropriately, let alone creatively and humorously, to Teuber's *Are you stiff?*

Consistent with this analysis, interpreting Teuber's *stiff* to mean *inebriated person* is the antithesis of *creative*. It is ungrammatical and disrupts ongoing communication in the context, *Are you stiff after this long drive?* To see this, replace *stiff* in Teuber's question with *inebriated person* and note the ungrammatical result: *Are you* [inebriated person] *after this long drive?* Did Henry really entertain that impossible interpretation and decide to go with it anyway? Not likely. Henry almost certainly did not understand either way of interpreting Teuber's question. Recall that in 1974, Henry detected double meanings in ambiguous sentences at below chance levels of accuracy.[18]

Labeling Henry's remark a *quip* also fits poorly with available data. Quip by definition means quick, but Henry was abnormally slow at comprehending ambiguous sentences—regardless of whether his comprehension was correct or incorrect.[19]

Extensive data also challenge the idea that Henry imagined that Teuber might enjoy talking about inebriation. Amnesics are especially poor at imagining.[20]

Available data even challenge Teuber's idea that Henry *deliberately* switched the topic of conversation to booze. Patients with hippocampal damage, including Henry, lack that kind of initiative. They are generally passive and reactive rather than proactive and assertive.[21]

Equally problematic for the creative quip story, a simple, low-level hypothesis with independent support readily explains Henry's *not a drop.* Tired after his long drive, Teuber may actually have said, *Are you . . . uh . . . stiff?* which Henry misheard as *Are you a stiff?* Without an audio-record of their conversation, there was no way to check this possibility. However, even without an . . . *uh*, Henry probably processed *stiff* as an isolated word, independent of its context. This would explain why the *inebriated person* idea was incompatible with the rest of Teuber's question. In well-controlled studies, Henry had difficulty integrating word-meanings into their novel context in a sentence (see Chapter 8).

UNEXPECTED JOLTS

Unexpected jolts make people laugh,[22] and quirky shocks abounded in Henry's social exchanges. My first shock came as Henry and I climbed the stairs to our testing room on the day we met. Henry said he collected guns. I was astounded. It seemed incredible that a profound amnesic could have access to firearms. Henry's inability to correct obvious errors in later studies also shocked me again and again. How Henry described one cartoon illustrates one such shock.[23]

Henry saw nothing funny in the cartoon. So the experimenter, Dr. Lori James, asked Henry to read its caption aloud, expecting him to say, *No, Thursday's out. How about never? Is never good for you?* Instead Henry said: "Yeah . . . uh, and so, just what he said, over the telephone. To the person he's talking to . . ." That was surprising. Henry never did read the caption aloud. He clearly did not understand it. However, the real jolt came when Henry continued, "And in, and—he's making double correction."

Puzzled, Dr. James asked, "He's doing WHAT?" "He's making a double correction," repeated Henry, ". . . because, 'it's never good for you,' means that, that never been good for the person he's talking—person he's talking to. And, he has stated something, he stated it about, person who's out, and he's just statin' it, the other person always ma—said that, uh . . . said it was never good. And he's just repeatin' something."

Whereas most native speakers of English understand what they are saying, Henry clearly did not know what he meant by *making a double correction.* His fabricated explanation shocked and saddened us. His *making a double correction* was just another of his unintended, uncorrected, and uncorrectable mistakes.[24] Listeners in casual conversation with Henry could only discover that by asking, *He's doing WHAT?* They did not.

Genuine humor also jolts the listener, but it satisfies in the end because the initial shock gets resolved. One is surprised when *re-word* replaces the expected *reward* in *A pun is its own re-word* but when its deeper meaning finally dawns—that this is a fun pun that illustrates what puns are—*re-word* is rewarding rather than silly and inappropriate. *Weak* instead of the anticipated *week* in *Seven days without a pun makes one weak* likewise might seem foolish at first, but it is fun-sense rather than non-sense. It achieves a pithy Hollywood ending and conveys a message about humor and health that is both appropriate and effective (see Reflection Box 23.2).[25] However, Henry's uncorrected and uncorrectable errors only shocked. Unlike real humor, they disturbed rather than satisfied in the end. They were not funny.[26]

REFLECTION BOX 23.2: BEWARE THESE WARNING SIGNS OF HEALTHY AGING[27]

1) ***Your sense of humor is unquenchable***.
2) Your outlook is *optimistic*, and you tend to frame events in a positive light.
3) You often experience *joy and happiness.*
4) You have a chronic tendency to *adapt well to changing circumstances.*
5) You rapidly *recover from challenges and stress.*
6) *You* love *physical activity.*
7) You enjoy *detecting and communicating feelings*.

8) You are *habitually generous.*
9) You enjoy an ongoing *support network.*
10) You frequently *express gratitude.*

THE ANECDOTAL EVIDENCE MYSTERY

How did Henry acquire his undeserved reputation for humor among well-trained and well-intentioned scientists schooled in the pitfalls of anecdotal evidence? They were not thinking scientifically when they accepted those Henry-stories at face value and passed them on. Believing that Henry's language skills were intact, scientists encountering Henry at MIT would not expect him to produce errors resembling *double correction* that he could not correct. Perhaps they misconstrued such astonishing Henry-isms as intended and witty! These hypotheses are untestable at this point. However, self-correction is the essence of science. Henry's make-believe wit can be credited with inspiring the first experimental investigation of humor and amnesia. Future research building on its results might eventually pay back on Henry's contract with science—the promise that what scientists learned about Henry will help others. The next chapter discusses how research with Henry did indeed help others understand and deal with memory loss.

QUESTIONS FOR REFLECTION: CHAPTER 23

1) Think about Henry's response when trying to explain the "The Pill! The Pill!" cartoon. Why do you think amnesics like Henry often confabulate and pretend to understand when they don't?
2) How has humor contributed to your life in the past? Give specific examples.**
3) Have you generated humor to help the lives of others? Give specific examples.**
4) Has your sense of what is funny changed as you've gotten older? How?**

5) If you were present during The "Hi, John" scenario, The Puzzle King scenario, and The Long Drive scenario, what two questions would you ask Henry to determine conclusively whether he intended to be funny?**
6) Henry probably experienced many funny cartoons and jokes before his operation. What if, after his operation, you had shown him his favorite cartoon or joke. Do you think he could explain what was funny? What might this show about Henry's sense of humor?**
7) Tongue in cheek, Reflection Box 23.2 presents "The Top Ten Warning Signs of Healthy Aging." Can you add other warning signs to this list? Which and why?**

TEST YOUR MEMORY FOR CHAPTER 23

1) What seemed to prevent Henry from comprehending novel cartoons?
2) What are some of the health benefits (physical and mental) of humor and laughter?
3) Why might the scientific analysis of humor resemble dissecting a frog?
4) How many warning signs of healthy aging can you remember? Would you remove any from the list?**
5) Why was it so easy for scientists schooled in the pitfalls of anecdotal evidence to credit Henry with a keen sense of humor?

MEMORY TEST ANSWERS: CHAPTER 23

1) pp. 296–97
2) Reflection Box 23.1, pp. 297–98
3) p. 300
4) Reflection Box 23.2, pp. 304–305
5) p. 305

SECTION V
COMPENSATING FOR CATASTROPHE

Chapter 24
PLAY IT AGAIN, HENRY

CAN EXPERTISE DEVELOPED OVER YOUR LIFETIME SHIELD YOU FROM NORMAL COGNITIVE DECLINE?

In 1970, Dr. Hans-Lukas Teuber made an important announcement. *Henry could gradually learn new information through repeated experience*! This was how Professor Teuber expressed this fact in a three-way conversation with his graduate student, William Marslen-Wilson, and Henry himself:

> **Hans-Lukas Teuber**: These things that he repeats, or that do repeat many times for him ... do get to him ... It's not all that bad, you know... it's... er... these last few years we did notice a real improvement, in the way you [talking to Henry] can recall things...[1]

In the 1960s, two high-profile events did indeed get to Henry: the televised assassinations of Jack and Robert Kennedy. Nonstop replays of those murder scenes apparently burned a long-lasting connection between *Kennedy* and *death* into Henry's brain.

Was *massive* repetition always necessary, I wondered? Could Henry create some types of new internal representations with a single repetition? In the follow-up to my 1966 ambiguity detection experiment, I noticed that Henry often echoed the last few words in sentences he heard.[2] Here an experimenter is explaining to Henry what this sentence means: *Mary and I approved of his cooking*. Note how Henry repeated the final words of her explanation.

Experimenter: Do you understand? That they approved . . .
Henry [echoing]: . . . that they approved, they agreed, they liked it.

Henry's spontaneous parroting was unusual. No control participant did that. Were Henry's immediate replays helping him create internal representations of sentence-meanings? This idea seemed consistent with other facts. During everyday activities, Henry often repeated random sentences over and over, as if trying to shore up their meaning in memory. The next day he would spontaneously repeat other sentences he had heard.[3] Henry even repeated whole stories virtually word-for-word, but to understand why, I needed more and better data.

The transcript of my ambiguity-detection experiment deepened the puzzle. When trying to describe the two meanings of ambiguous sentences, Henry frequently repeated the very words he was supposed to clarify: the *ambiguous* words.[4] This is Henry explaining the two meanings of *The marine captain liked his new position*. Note his repetitions of the ambiguous word *position*. In all, Henry repeated *position* five times without ever explaining its other meaning in this sentence:

> **Henry**: The first thing I thought of was a marine captain he liked the new *position* on a boat that he was in charge of, the size and kind it was and that he was just made a marine captain and that's why he liked the *position*, too, because he was above them and of all, most of all . . .

Henry repeated the ambiguous words in ambiguous sentences almost three times as often as carefully matched control participants—which was remarkable because the experimenter again and again asked Henry to stop doing that and explicitly requested a different meaning for the ambiguous words he was repeating. It was as if repetition was the only way Henry could connect the meanings of the ambiguous words with adjacent concepts to create a coherent internal representation of the sentence.[5]

Henry also repeated *unambiguous* words in the sentences, albeit considerably less often than the *ambiguous* words. Note in this excerpt Henry's seven repetitions of the unambiguous words *stay* and *home* (in bold) in the sentence, *The stout major's wife stayed home*:[6]

> **Henry**: She ***stayed home***, she ***stayed home*** or was not moving around . . . Then, uh, sort of, or made to, or *to* ***stay*** at ***home*** was *to* ***stay***, not go out, not leave. . . [repeated words are in italics]

Note also Henry's focus on isolated words. His *was not moving around* accurately defines the word *stayed* in isolation but not in the phrase *stayed home.* Being *immobile* and *staying home* are not synonymous. This focus on isolated words often undermined Henry's attempts to link word-meanings with their novel context in a sentence. Despite his repetitions, Henry did not understand what most of the sentences meant. Did Henry require *massive* repetition to form new and *accurate* internal representations after his lesion? Again I needed better data to determine whether my mentor was right about things that "repeat many times" for Henry.

ELABORATIVE REPETITION, ONE CONNECTION AT A TIME

I found what I wanted in Henry's picture descriptions on the Test of Language Competence.[7] This is Henry describing a TLC picture of two high-school students in conversation. The one shown speaking is pointing at an overcrowded school bus:

> **Henry**: . . . she wants to go on the bus . . . and it's crowded . . . *it's crowded* . . . too *crowded* to get *on the bus* [repeated words are in italics].

Note how Henry twice repeated the phrase *it's crowded* before achieving what he wanted to say: *it's too crowded to get on the bus.* He first conjoined *it's* with *crowded,* and then linked *too* with *crowded* to form *it's too crowded.* That pattern—*repetition with elaboration*—appeared over and over in Henry's TLC descriptions and was the only type of repetition that Henry produced abnormally often. Immediately reiterated words (*this . . . this*), speech sounds (*s- school*) and phrases (*but it was, but it was*) were as common for memory-normal controls as for Henry. The most plausible explanation is that Henry's *elaborative repetitions* reflect a deliberate strategy to offset his problems in forming new internal representations. By producing a familiar word or phrase and then internally or overtly repeating it with elaboration, Henry was able to form internal representations for novel phrase-level plans via repetition, one link at a time.[8] Henry had discovered how to create simple new connections via elaborative repetition. As discussed shortly, anyone trying to learn someone's name can benefit from Henry's elaborative repetition strategy.

So, if *massive* repetition was necessary for Henry to represent events such as the Kennedy assassinations, it was unnecessary for creating new internal representations (memories) for phrases. Despite his hippocampal damage, Henry required very few repetitions to form new internal representations for the phrase *too crowded to get on the bus.*

Why was it relatively easy for Henry to create memories for new *phrases*? As one contributing factor, Henry's internal representations for phrases were *temporary*. They only needed to survive *briefly* in memory—just long enough to get out the door on the TLC. If asked to describe the same picture the next day, Henry would almost certainly require another bout of elaborative repetition to re-create the phrase *too crowded to get on the bus*—plus many additional repetitions should he want to transform that phrase into a relatively permanent fixture in his brain.

Additional contributing factors were the knowledge Henry acquired as a child about English grammar and the part-of-speech of familiar words. Part of speech determines how to conjoin words into grammatical phrases. Henry knew that *too* is an adverb and that *crowded* functions as an adjective in the expression, *It's crowded*. By internally repeating *too* and *crowded*, Henry could therefore form an internal representation for the concept *too crowded* and ultimately for the idea that he wanted to express: *It's too crowded . . . to get on the bus.* The knowledge that *crowded* functions as an adjective prevented Henry from saying *It's crowded . . . too*, rendering ungrammatical any possible completion of this sentence.[9]

In the normal brain, the hippocampus speeds up the process, rendering overt, *behavioral* repetitions unnecessary. However, part-of-speech knowledge specifying what to conjoin with what also helps the normal hippocampus create one new connection after another—and *quickly*.[10]

ARBITRARY VERSUS STRUCTURED ASSOCIATIONS

Henry and normal older adults experience special difficulty in learning and remembering *arbitrary* associations between unrelated information, for example, random word-pairs such as *death-table*. The underlying basis for this difficulty is *ambiguity*: most common words are ambiguous *in isolation*. The noun meanings of *table* include a piece of furniture, a negotiating session (he brought the idea to the bargaining *table*), a systematic arrangement of data (the *table* shows the salary of each employee), a summary enumeration (a *table* of contents), a group of people assembled at a (possibly metaphoric) table (he had the attention of the entire *table*), things with a plane surface (we are in *tableland*), up for consideration (that topic is not on the *table*), into a stupor (she drank him under the *table*), in a covert manner (he took money under the *table*). Then come the many *verb* meanings of *table*. . . . Because isolated words are so ambiguous, no well-practiced rules are available for specifying how to integrate the meanings of random pairs of isolated words. Perhaps my mentor was right that *massive* repetition was the only way Henry could form *permanent* links between *disconnected* concepts.

Even *structured* associations built into the sentences and paragraphs in books, newspapers, and magazine articles were difficult for Henry to comprehend and remember. Capturing the gist of prose passages is not as simple as combining words into phrases such as . . . *too crowded*. To create a memorable internal representation of a story, one must quickly select, organize, and manipulate the incoming information. This Henry could not do.

The *structured* concepts in novel stories are even difficult for normal older adults to learn and remember. Beginning around age fifty, we gradually retain less and less information from prose passages as we grow older.

However, Professor Art Shimamura and a team of neuroscientists at the University of California, Berkeley discovered an important exception to that *age fifty* rule. Despite suffering normal age-linked declines in processing *disconnected* concepts, *older professors* at UC Berkeley could recall lengthy prose passages as readily as sharp young professors at the same university, and *much better than non-professors aged fifty and older*.[11]

To show this, Dr. Shimamura recruited five groups of participants: the standard comparison groups of young and older *non-professors*, plus three groups of unretired professors with similar education in a wide range of academic fields: *young* professors (average age thirty-eight), *middle-aged* professors (average age fifty-two) and *older* professors (average age sixty-five).

The researchers first showed that aging impaired the basic mechanics of cognitive function—reaction time and the brute force association of unrelated concepts—to the same degree for professors and non-professors. For both groups, aging reduced recall of unrelated words and slowed reactions to simple stimuli. For example, if participants studied a list containing the word-pair *death-table* and later saw the word *death*, young participants correctly recalled *table* much more often than the older ones.

The researchers next asked participants to remember lengthy passages of prose with no relation to their area of expertise: fictional stories about everyday life and factual narratives from the worlds of science and history. To everyone's surprise, the senior professors recalled as many concepts from the stories as the young professors, and many more than the older non-professors, regardless of the type of concept.[12] Skills developed on a daily basis throughout their careers helped the older professors link the new knowledge in the stories with existing representations in their brain. They were able to efficiently select, organize, and manipulate new information to create memorable internal representations—and thereby compensate for the age-linked damage to their basic biology that slowed their reaction times and hindered their acquisition of disconnected concepts. The knowledge and skills you develop in life can shield you from otherwise normal cognitive decline.[13]

TO LEARN ARBITRARY ASSOCIATIONS, DON'T JUST RE-ACTIVATE, ELABORATE

Henry's strategy of combining *elaboration* with *repetition* is a very effective way to form face-name links—the most important type of arbitrary association that people form in everyday life. The idea is to create an elaborate network of unique links between the face, the name, and the many related images and concepts that already exist in your brain. Then strengthen those new links via repeated activation.

To do this, ask as many questions as possible about the person and their name when you meet them. *Glad to meet you, Tryna MacArthur. Is your name spelled Tryna or Trina? Did that spelling originate in Eastern Europe? Is Tryna short for Katryna? Is your surname spelled with the Scottish Mac or the Irish Mc? How long have you been a journalist, Tryna? What is your area of specialization? May I please have your business card?* Be prepared of course to answer similar questions about your own name and personal history.

Meanwhile, study Tryna's face, smile, hairstyle, clothes, body language, voice quality, and accent. Seek out unusual or distinctive features that might make her easy to remember. Then link Tryna with her name in as many ways as possible. Is Tryna a high-energy person? Create an image of Tryna battling Hurricane "Katryna" in New Orleans. Or imagine journalist Tryna *MacArthur* interviewing General Douglas *MacArthur* in Tokyo on his mission to end World War II. Then repeatedly retrieve your images. Repeated activation of unique and vivid images is the best way to permanently cement a face-name link in memory. The next chapter will examine name learning in greater detail, especially the surprising relation between names and the binding mechanisms in the hippocampus that research with Henry revealed.

QUESTIONS FOR REFLECTION: CHAPTER 24

1) Why must people repeat some types of information for accurate recall from memory whereas a single exposure suffices for other types of information? Can you think of some examples of both kinds and the types of contexts that distinguish them?**
2) This chapter described an important experiment comparing the ability of young and old professors versus old non-professors recalling "lengthy prose passages about information uncon-

nected to their area of expertise." In this study, older professors outperformed all non-professors and were on par with younger professors. How would you predict the results to change if the information to be learned was directly connected to area of expertise or personal experience of the participants?**

3) What knowledge and skills have you developed in your life that can "shield you from otherwise normal cognitive decline"?**
4) Why do you think that reaction time (the speed of responding to a stimulus) slows down with aging?

TEST YOUR MEMORY FOR CHAPTER 24

1) What is "repetition with elaboration"? What is the likely reason Henry adopted that strategy?
2) How does elaboration help retain the link between names and faces in memory?
3) How does the hippocampus facilitate the learning of arbitrary associations between unrelated information, such as random word pairs like *death-table*?
4) What are some tricks for remembering someone's name when you meet them?
5) Which words did Henry repeat more often in Dr. MacKay's ambiguity detection experiment, ambiguous or unambiguous? What was the apparent reason for this difference?

MEMORY TEST ANSWERS: CHAPTER 24

1) pp. 311–12
2) p. 315
3) pp. 312–13
4) p. 315
5) p. 313

Chapter 25

DO YOU REMEMBER WHAT'S-HER-NAME, HENRY?

WHAT IS YOUR BEST STRATEGY FOR ENSURING THAT PEOPLE REMEMBER YOUR NAME?

Face-name pairings are hard to learn and remember, especially for older adults. I discovered these facts by reading the scientific literature on the *baker/Baker* paradox. Here is the first enigmatic result—replicated in laboratories around the world.[1] One group of students in a memory experiment sees a photo of a face and learns she is a *baker*. Another group sees the same photo and learns that her last name is *Baker.* Later, both groups revisit the photo and the *baker* group tries to recall her occupation while the *Baker* group tries to recall her name. Same photograph. Same speech sounds. Different recall. The *baker* group answers correctly more often than the *Baker* group. Why is that?

The cognitive aging lab of Dr. Lori James repeated this *baker/Baker* experiment with older adults. Now the *baker* group answers correctly *much more often* than the *Baker* group. The superior recall of *baker* versus *Baker* has grown with aging. Why is that?

Psychologists solved the *baker/Baker* paradox in the 1990s. When participants learn that the woman in the photo is *a baker*, they link their internal representation of her face with an already formed set of ideas about the typical baker. They might imagine her in a baker's hat, wearing a white apron, baking bread at four a.m., circles under her eyes. Perhaps they hear her complain in their mind's ear as she drives home from a hard night's work at the local bakery, smelling of yeast, with a splotch of flour in her hair. These vivid images link her face to her occupation via an array of connections. When the experimenter later revisits that photograph and asks, *What's her occupation?*, those diverse pathways leading from face to *baker* allow a fast and accurate response!

Not so with *Baker*. When participants see a snapshot of someone named *Baker*, no existing network of ideas about the name *Baker* can help connect *Baker* with her face. They are forced to link *Baker* to their internal representation of her face via a single connection that can easily become defunct due to aging, non-recent use, and infrequent use over their lifetime—rendering *Baker* impossible to remember in the subsequent memory test. The superior recall of *baker* versus *Baker* therefore grows in magnitude with aging because degradation is especially likely to disable that one connection in older adults.

HOW TO LEARN A THOUSAND PEOPLE'S NAMES AS AN OLDER ADULT

My understanding of the *baker/Baker* paradox shifted from *intellectual* to *personal* when I tried to learn around a thousand names as an older adult. Most of the names belonged to students in the smaller classes I taught in the final decades of my teaching career at UCLA. The maximum enrollment for my *Laboratory in Cognitive Psychology* course was twenty-four students. We devoted 99 percent of our class-time to cutting-edge science and one percent to a collaborative learning project. The goal of the project was for everyone in the class, myself included, to learn the first and last names of everyone else.

Learning those names galvanized the class. This is how we did it. On the first day of the course, I promised I would learn the students' names. I just needed their help. I asked them to correct me if I mispronounced a name. People learn by correcting their mistakes. I also asked for patience. "Expect me to be slow," I said. "Older adults learn and retrieve newly encountered names much more slowly than young people."

I then divided the class into six research-presentation teams and snapped a group photo of each team. Not police mugshots. Each student held a large placard displaying their full name in a colorful font and hammed it up for the camera like actors from nearby Hollywood.

The following week, each team gave their first brief presentation. The goal was to teach the class something distinctive and interesting about each team-member, including their first and last name. To reward effective teaching and learning, an extra-credit question on the midterm might test recall of a student's name.

To help with the assignment, I illustrated how to transform hard-to-learn *Bakers* into easy-to-learn *bakers* using my own name. "Begin your presentation by firmly establishing in everyone's mind the spelling, origin, and pronunciation of your name," I advised. On a whiteboard I printed *Don* followed by the ancient Scottish surname, *MacKay*. I noted that "*Mac* means *son of*," whereas the meaning and origin of *Kay* is a mystery. If *Kay* is short for *Katherine*, the name *MacKay* is unusual. The historic root of most Scottish surnames is the given name of a *male* chieftain, as in *MacDonald*, *MacArthur*, and *MacHenry*. *MacKay as* short for *MacKatherine* would be a whodunit.

The Scottish pronunciation of *MacKay* is also curious. It rhymes with *eye* and *I*. To illustrate, I sang the chorus of a once-popular Scottish folk song: "*Oh*, I like MacKay and MacKay likes me." The students were amazed to hear a UCLA professor sing. "Surprise helps recall," I said. "So does rhythm and rhyme."

"Next, tell the class something about yourself and your name. Your story can be false and preposterous," I suggested, "as long as it leaves an unforgettable trace in everyone's mind."

To illustrate, I asked the students to imagine me on a kayaking expedition departing at dawn for the Khyber Pass. "I'm wearing a kilt. My kayak is loaded with Big Macs. The resulting image will help you remember my SCOTTISH ANCESTRY (the kilt) and my names: DAWN (*for Don*) + *MAC* (*Big Macs*) + *KAY* (pronounced as in *KAY•yak* and *KHY•ber*). Distinctive and dramatic visual images are especially memorable," I emphasized. "Names are easier to forget than images."

"Non-fiction can also be memorable," I added. "This is a true story about *MacKays* emigrating from Scotland to North America. They mainly landed in *Nova Scotia*, which means *New Scotland*. That's where my family settled. For almost two hundred years, we *MacKays* have corrected American immigrants to Canada who think that *MacKay* rhymes with *gay* and *hay*. That mispronunciation and its tell-tale spelling, *McKay*, came via *Ireland* to America."

"True stories based on personal experience carry extra oomph," I suggested. "Here's one. In his final year of college, my son wanted to explore his roots. I suggested he read *The MacKays in Canada*, an unpublished monograph written by his grandfather. 'I prefer first-hand experience,' he insisted. 'Let's do a father-son vacation in the Scottish highlands.'

"We flew to Inverness that summer. While exploring the spectacular north coast of Scotland by car, we discovered by chance *The MacKay Museum* in the tiny village of Strathnaver. It was a two-story, recycled church containing everything known, imagined, or associated with the *MacKay Clan*, including a newspaper article about *The MacKays in Canada*. A photograph showed its chieftain wearing a kilt, waving a scepter, and leading a parade of bagpipers down the main street of Toronto. It was my father. 'I believe I have found my roots,' was my son's understatement."

After assuring the students that my tricks for transforming *Bakers into bakers* would work with other types of names, whether German, Chinese, Zimbabwean, Hungarian, Armenian, or Russian in origin, I reviewed the two ways to create a memorable face-name link in memory. "My true and invented stories conveyed similar information," I noted, "but for the true versions, *you*, the listener, actively constructed the links between me, my Scottish heritage, and the pronunciation of my name, whereas *I* created those links in my fictitious story. The difference is important. Research shows that knowledge is better remembered when self-generated rather than passively received."[2]

The students loved the assignment. Everyone conveyed memorable information about their names and quickly learned the names of everyone else. In subsequent years, only one student mispronounced my name. She had missed the first class.

The students also enjoyed the effects of their presentations. When I managed to recall a challenging name in class, I raised my arms like a triumphant soccer player. The students also celebrated. For many, I was the only professor to learn their names after four years at UCLA. They did not care that the secret to my success was the team photos I had shot. By installing those snapshots on my computer and revisiting them occasionally, I was able to conjoin their names and faces in my brain using every effective technique developed since Ebbinghaus for enhancing memory—*semantic elaboration* of the pictures based on information in the student presentations, plus *expanding rehearsal*—where I gradually increased the interval between studying a face and testing my ability to recall the name. As an older adult, I needed all the help I could get.[3]

A PROPER NAME MYSTERY RESOLVED

When it came to names, Henry was oddly normal. Despite his profound inability to learn other types of new information after 1953, Henry learned and remembered proper names just like normal people—a weird fact that surfaced over and over in William Marslen-Wilson's 1970 interviews.[4] Proper names that Henry had no business knowing kept popping up in his answers to Marslen-Wilson's questions: *Robert Kennedy*, *John F. Kennedy*, *Lee Harvey Oswald*, *Richard Nixon*, *Marilyn Monroe*, *Barry Goldwater*, *Billie Jean King*, *Martin Luther King*, *Lyndon Johnson*, *Pat Nixon*.[5] These names formed part one of a three-part mystery. Henry must have learned them after his 1953 operation, but without a hippocampus, how could he conjoin the first name *Lee*, with the middle name *Harvey*, and the last name *Oswald*?

As part two of the mystery, Henry linked new names with new facts featured in the news after his surgery:

> **Marslen-Wilson**: Do you know anything about a war in Vietnam?
> **Henry**: ... In a way I don't ... know the ... anything about it in a way ... but ... uh ... Americans ... went over to help ... fight over there.[6]

Did massive repetition burn that link between Vietnam and American help into Henry's brain? Perhaps. However, Henry also learned *obscure* facts about *low-profile* proper names confronted after his operation. Did Henry encounter the name *Yoko Ono* often enough to burn a new connection between *Yoko Ono* and *The Beatles* into his brain? It seemed unlikely.[7]

As part three of the mystery, Henry detected and corrected self-produced errors involving proper names, as in this excerpt from his 1970 interviews. Marslen-Wilson had just asked Henry to name the leader of the North Vietnamese forces:[8]

Henry: . . . and . . . I think of Shek right off . . .
Marslen-Wilson: Shek?
Henry: Chiang Kai Shek.
Marslen-Wilson: Chiang Kai Shek.
Henry: That's right . . . Chiang Kai Shek . . .
Marslen-Wilson: You think the Americans are fighting against him in Vietnam?
Henry: . . . and . . . uh . . . Vietnam is . . . uh . . . not . . . uh . . . part of . . . uh . . . well it's . . . in Asia but not part of China.
Marslen-Wilson: No, that's right . . .
Henry: And . . . uh . . . I believe he . . . uh . . . uh . . . I believe the Americans are fighting against the Soviet Union . . .
Marslen-Wilson: Where?
Henry: In Chiang Kai Shek . . . uh . . . not Chiang Kai Shek but the . . . uh . . . well . . . Vietnam.

Here Henry had intended to say *Vietnam*, but said *Chiang Kai Shek* by mistake. Nothing unusual there. Name-substitution errors are common in normal people. However, Henry also *detected* his error, *corrected* it, and introduced the error-markers *uh*, *not*, and . . . *well* to signal its occurrence. How remarkable! Henry had produced hundreds of errors involving other types of words, but *detected, corrected and marked none of them*.[9]

What was it about Henry's brain that made names special? How could Henry form the novel internal representations required to detect and correct errors involving proper names but not other categories of words? Did his 1953 surgery spare his binding mechanisms for proper names? *This* could explain all three aspects of the proper name mystery: Henry's ability to learn new name sequences like *Lee Harvey Oswald*, link them with novel facts, and correct his self-produced errors involving proper names—just like a memory-normal person.[10]

Henry's operation destroyed only about half of his hippocampus, so it was not entirely implausible that some binding mechanisms in his hippocampal region might be intact. But how could I test that idea? Again I turned to Henry's responses on the Test of Language Competence.[11] Again I found exactly the evidence I needed.

IMAGINARY NAMES TO THE RESCUE

The data showed that Henry *invented* proper names to describe people in TLC pictures in order to compensate for his difficulties in describing them via other means. In a wasteland of broken binding mechanisms, Henry's mechanisms for conjoining proper names were working to save the day.

Henry named the strangers in TLC pictures *David*, *Gary*, *Jay*, *Melanie*, or *Mary*. This was extraordinary. Normal participants always described TLC people using pronouns and phrases such as *this woman*. Why didn't Henry do the same?

The answer was not hard to find. As the following excerpt illustrates, Henry had difficulty describing people via pronouns and common nouns of the appropriate gender. The TLC picture shows two men conversing about a third man climbing a steep cliff.

> **Henry**: *David* wanted him to fall and to see what lady's using to pull himself up besides his hands.

Henry's invented name, *David*, designates the TLC man shown speaking and pointing. No serious problem there. However, the common noun *lady* in Henry's sentence is the wrong gender for referring to the climber, and the pronouns *himself* and *him* are the wrong gender for referring to their antecedent in the sentence (*lady*). No control participant made errors like that. Without a hippocampus, Henry had difficulty creating internal representations containing gender-appropriate pronouns and common nouns.[12]

Henry's David excerpt also illustrates his struggles with noun phrases. For his sentence to qualify as grammatical under TLC rules, Henry should have said something like *to see what the lady's using.* For Henry to say *see what lady's using* he must have had difficulty creating *the lady*, the simplest possible noun phrase containing the common noun *lady.*

However, no similar difficulties marred Henry's use of *proper* nouns. He always created gender-appropriate names for the women and men depicted in the TLC. It wasn't that Henry named TLC people after acquaintances from before his operation. The names Henry invented on the TLC matched none of the dozens of names Henry mentioned in his 1970 interviews with Marslen-Wilson. Henry created his imaginary names afresh, consistent with the hypothesis that his mechanisms for linking a person to a gender-appropriate proper name were intact.[13]

TESTING THE COMPENSATION HYPOTHESIS

Did Henry invent proper names to compensate for his inability to create novel sentences containing gender-appropriate pronouns and common nouns in everyday life and elsewhere? To test this idea, my lab examined how Henry and a comparable group of memory-normal individuals answered questions about childhood experiences. Although none of the questions called for use of a proper name, Henry included significantly more proper names in his answers than did the controls. In the excerpts below, Henry and a typical control participant are answering the question, *What is your first memory, the earliest thing you remember?* Note that the control participant included no proper names whereas Henry used four: *Hartford, Manchester, South Coventry*, and *Burnside.*[14]

Henry: When I . . . tell you that 'tis . . . you see . . . may have been . . . that was when I was going to high school . . . that . . . and . . . but before that when I was going to the private kindergarten, two houses up, from where I lived, when I went to high school, but the other places I lived in Hartford, and Manchester, and then South Coventry . . . before coming back to [chuckles] Burnside Avenue again.[15]

Typical Control Participant: Oh, way back, uh . . . two. I was two because I have seen pictures of myself in a snowsuit, and I outgrew it very quickly, but when I was two, I wore it, and when I was two, I remember walking in my grandma's kitchen and pointing up at my snowsuit hanging on the kitchen door because I wanted to put it on, and it's very clear—it was light blue.

Would Henry still use more proper names than comparable control participants if he did not have to maintain the question in memory? A subsequent experiment asked for *written* responses to *written* questions and Henry used even more proper names than before.[16] For example, when asked, *Can you describe any children in your grade school?* Henry wrote *MAN. S.P.S. HTFD.* (The punctuation and capital letters are Henry's.) When asked what those abbreviations meant, Henry indicated that *MAN.* stood for *Manchester*, *S.P.S.* for *Saint Peter's School*, and *HTFD* for *Hartford Fire Department.* This was astounding. Henry's description of grade school acquaintances consisted *entirely* of unrelated proper names! If Henry wanted to write something, *anything*—no matter how incoherent or irrelevant—proper names were his fallback. Henry's binding-mechanisms for proper names must have been an oasis in the desert of shattered mechanisms in his hippocampus for creating new memories.

Proper names were by far the most carefully mapped survivor of the tragic damage to Henry's binding-mechanisms. However, other categories of binding-mechanisms were almost certainly spared as well. One was ongoing topics of discussion. An earlier chapter illustrated how Henry *inadvertently* spiraled his conversations into loops, circling back to an original topic of conversation after nearly twenty speaker-turns discussing unrelated issues. Henry also *intentionally* recalled topics after remarkably long dialogue-filled intervals, as in this excerpt that immediately follows Henry's earlier comment about Americans "helping out" in Vietnam:[17]

Marslen-Wilson: Uh-huh ... right ... um ... How are you feeling? ... Tired?

Henry: Well ... I'm just wondering myself ... now ... well ... when you fellows are taking this all down of course on tape ... but I'm wondering just how it will be ...

Marslen-Wilson: How do you mean, "how it will be"?

Henry: Well ... just how I have spoken, how I sound, and what my answers are ... and ... uh ... a a big question mark right there ...

Marslen-Wilson: Your answers are very helpful, very helpful indeed.

Henry: They are ... I hope so ...

Marslen-Wilson: But are you beginning to feel tired perhaps ...?

Henry: Um?

Marslen-Wilson: Are you ... are you at all tired? We've asked you a lot of questions.

Henry: No, I'm not tired.

Marslen-Wilson: Do you know what the time is?

Henry: Well, by this it says ... twenty minutes of four [pointing at a clock].

Marslen-Wilson: Ah ... well, what have you ... what have we been asking you questions about?

Henry: Well about ... the war, Chiang Kai Shek ... the war ... China ...

Marslen-Wilson: Before that?

Henry: And Indochina ... and ... about us ... uh ... helping out ...

This excerpt was astonishing! In response to explicit questions, Henry correctly recalled five prior topics of conversation, including one mentioned twenty-six speaker-turns earlier (Americans helping out in Vietnam), an interval filled with 133 words of irrelevant discussion about what time it was, whether he was tired, and how he was helping science. How did Henry do that with a broken hippocampus?

In 1970, most researchers assumed that Henry was "marooned in the present" and unable to remember any type of novel experience for longer than a few seconds.[18] Henry's ability to recall novel topics of conversation after much longer intervals clearly contradicts that assumption. The simplest explanation for this impressive feat of recall is that Henry's operation spared his binding mechanisms for representing topics of conversation as personally experienced events occurring at a particular time and place.[19]

Topics of conversation are of course just one of many different types of event that ordinary folks quickly and easily commit to memory. First and last names likewise represent only two of perhaps hundreds of categories of information that normal speakers effortlessly conjoin when producing sentences. How many special binding mechanisms does the normal hippocampus contain? Without Henry's help in understanding his memory, mind, and brain over the past fifty years, nobody could have asked, let alone begun to answer, that question.

QUESTIONS FOR REFLECTION: CHAPTER 25

1) Imagine you are a participant in the *baker/Baker* study. How could you break the paradox and recall that someone is named *Baker* as easily as recalling that they are a *baker*?**
2) Why do you think you might better remember information that you self-generated rather than passively received?**
3) Dr. *MacKay* invented a story about a *kay*ak filled with Big *Mac*s headed toward the *Khyber* Pass at *dawn* to help his students remember the name *Don MacKay*. What creative story would you tell to help people remember your first and last names?**
4) Imagine two people learning a name: one focuses on visual aspects of the letters in the name (the orthography), and the other focuses on auditory aspects of the sounds in the name (the phonology). Which one do you think will better remember the link between face and name?**
5) In this chapter, Dr. MacKay discusses creative strategies for helping his students learn the names of everyone in the class. Can you think of some additional techniques to improve name recall in your life?**

6) What does Henry's preserved ability to learn and use proper names tell us about how the brain handles common versus proper nouns?

TEST YOUR MEMORY FOR CHAPTER 25

1) What is the *baker/Baker* paradox and how do you explain it?
2) How might the learning strategies known as *semantic elaboration* and *expanding rehearsal* help you remember the link between a face and a name?**
3) What is the significance of Henry's ability to create proper names with the correct gender despite his inability to use pronouns with the correct gender? What does this say about how the brain and hippocampus work?
4) How did Henry compensate for his difficulties in binding pronouns to common nouns and in creating novel phrases containing common nouns?
5) It was previously assumed that Henry was "marooned in the present" (unable to retain novel experience for longer than a few seconds). What evidence discussed in this chapter indicates that idea was *not* correct? (hint: topics of conversation)

MEMORY-TEST ANSWERS FOR CHAPTER 25

1) pp. 319–20
2) pp. 323–24
3) pp. 325–26
4) p. 327
5) p. 328

SECTION VI

CELEBRATION AND COMMEMORATION

Chapter 26
LET'S CELEBRATE A PROMISE FULFILLED, HENRY

HOW CAN YOU BEST MAINTAIN YOUR MEMORY, MIND, AND BRAIN?

Do you remember Samantha Kimmey? She was the news reporter who interviewed me in 2014 about my research with Henry. Chapter 1 summarized that interview to set the stage for *Remembering*. I recently exchanged emails with Samantha. After writing all but this last chapter, I wanted to thank her for asking the question that launched this book: how could my research with Henry help readers of her Pulitzer Prize–winning newspaper understand and maintain their memory, mind and brain? Now Samantha had a new question: "How does *Remembering* go beyond our interview of four years earlier?" This chapter summarizes my response.

A FOCUS ON HELPING HENRY HELP OTHERS

Remembering is about making good on a promise. In return for helping science understand his memory problems, scientists promised Henry that what they learned about him would help others. Marslen-Wilson expressed that verbal contract in 1970, and Henry reiterated it for the rest of his life. To honor that promise, *Remembering* shows how insights derived from research with Henry could help anyone maintain their memory, mind, and brain.

Henry's search for meaning in life illustrates an insight that combines the personal with the scientific. Making life worth living was difficult for Henry after his 1953 surgery devastated his ability to comprehend and remember new information. Unlike ordinary people, Henry could not find existential meaning in great works of art, music, literature, history, or philosophy. Quite the opposite. It must have been frustrating for Henry to hear a classic story or to gaze on a beautiful painting and be unable to understand or remember it.

True, Henry did not constantly struggle to survive after his surgery. Medicaid and Social Security covered his basic food, shelter, security, and medical needs at the Bickford Health Care Center. Private and federal grants also funded the VIP treatment Henry received during his more than fifty working vacations at the MIT Clinical Research Center.[1] However, we humans do not live by bread, comfort, and safety alone. Henry had no interest in sex, but how did he feel about children, another basic pleasure in life? Did the idea of having children of his own ever cross his mind? What did he feel about the dying of the light as he grew older? Nobody knows. These are just some of the questions I wish someone had asked Henry before it was too late.

Nonetheless, *Remembering* did settle one personal issue about Henry. He was clearly anxious and depressed in 1982. Contrary to Dr. Corkin's published claim that Henry's 1982 responses on the Beck Depression Inventory (BDI) revealed "no evidence of anxiety" or depression, my statistical analyses indicated that his BDI scores fell within the category *borderline clinical depression*—one level worse than *mild mood disturbance* and two levels worse than *normal ups and downs.* This finding suggests a possible explanation for Henry's hostility toward his mother and for his attacks on fellow residents at Bickford. Inasmuch as hostility and irritability are symptoms of depression, this also explained Henry's relative calmness at MIT. Henry's belief that he was helping scientists at MIT help others added meaning and purpose to his life, enhanced his self-image, and helped relieve the anxiety and depression contributing to his aggressive impulses.

BIG PERSPECTIVES

Remembering also provides some big perspectives on neuroscience. One is the contrast between the *brain-first approach* and the *feat-first approach*—the two fundamentally different frameworks that guided research with Henry from 1953 to the present. The brain-first approach focused first and foremost on Henry's brain damage, aiming to assign his impaired functions to the damaged regions of his brain and his unimpaired functions to the intact regions. Suzanne Corkin summarized that focus in her book: "The only way to truly understand the nature of Henry's amnesia would be to look at his brain (*post mortem*) and document the damage."[2]

Dr. Corkin's commitment to the brain-first perspective helped me understand her curious euphoria just hours after Henry died in 2008. Henry was her companion and protégé for more than forty years. However, Professor Corkin smiled, cheered, raised her hands overhead, and stood on a chair to applaud as the neurosurgeon in charge lifted Henry's brain from his skull and placed it in a metal bowl. Success of that procedure was critically important for Dr. Corkin, and at that moment, she was totally focused on it. At last, she could "learn with certainty which medial temporal-lobe structures were preserved in Henry's brain and to what extent"—the crowning achievement of the brain-first approach. Soon to come for Dr. Corkin were more celebrations, media attention, and a legal battle over the custody and final resting place of Henry's brain.[3]

I had no reason to celebrate when Henry died in 2008. Although I had lived and breathed the brain-first approach as an MIT graduate student, I adopted the feat-first approach after I left MIT in 1967. That feat-first approach focuses first and foremost on how the *normal* brain accomplishes the everyday feats that Henry's brain could not: understanding novel sentences, comprehending unfamiliar visual scenes, remembering new facts and events.

This focus seemed essential for fulfilling the promise that research with Henry would "help others"—a distant dream in 2008. I still had not unraveled the crossword puzzle mystery—the conflict between the anecdotes about Henry's ability to solve difficult crossword puzzles and my experimental findings indicating deficits in his word knowledge. In 2008, my data on Henry's ability to comprehend captioned cartoons were also meeting mountains of resistance from reviewers who seemed to prefer the widely circulated stories about Henry's "keen sense of humor." Even Henry's poor performance on the Test of Language Competence (TLC) was facing opposition. One cognitive scientist rejected in public and in principle my solid TLC experiments with Henry in favor of his personal impressions! The situation reminded me of Nobel laureate Daniel Kahneman's prophetic words about scientific reasoning: "I'm not very optimistic about people's ability to change the way they think."[4]

Remembering nevertheless challenges the intuitive way of thinking that labeled Henry a pure amnesic, someone with only one fundamental impairment: inability to remember new facts and events. Like other diagnostic categories in the *Diagnostic and Statistical Manual of Mental Disorders*, that *amnesia* label was supposed to stimulate research into the causes and possible remedies for a patient's impairments. That did not happen with Henry. *Remembering* shows how Henry's *amnesia* brand hindered research into his condition and prevented adoption of available remedial treatments that might have helped him.

TURNING POINTS

Remembering also highlights some historical turning points in my research with Henry. One occurred in the 1990s, when a team of Italian neurophysiologists discovered *mirror neurons*. These neurons *simultaneously* represent perception and action! They become activated *either* when someone performs a particular action *or* when they see someone else perform that action. The existence of mirror neurons challenges the way people thought about the brain from the 1700s to the present. Contrary to the philosophical framework of René Descartes, information processing does not always proceed in sequence from sensory perception, to comprehension and thought, followed by memory-storage, memory-retrieval, and finally the muscle-movements of action. One and the same neuron can achieve perception and action! This discovery helped explain the parallel deficits that my lab had observed in Henry's ability to comprehend, store, and produce novel linguistic information.

The year 2009 was another turning point. That was when my lab went public with Henry's deficits in comprehending novel visual information.[5] Just as with sentences, Henry understood familiar but not novel aspects of visual scenes. Demonstrating those visual cognition deficits was important. Three teams of British researchers had observed similar perceptual deficits in amnesics with hippocampal damage.[6] They too had experienced unwarranted rejection of their results. I was not alone. Their support for the idea that the hippocampal region contributes to creating new internal representations for visual as well as linguistic information has helped initiate a sea change. The brain-first approach will soon discover that there's more to memory than facts, events, and procedures.

CREATIVITY AND INSIGHT

Remembering also highlights topics that my 2014 interview with Samantha skipped. One is my serendipitous discovery of a relation between insight, creativity, and the hippocampus. In my 1966 ambiguity-detection experiment with Henry, the control participants often whispered *aha* right before they reported discovering a second meaning in ambiguous sentences—exactly what people say after achieving insight into a problem. Discovering the alternate meaning of ambiguous sentences must require creativity of some kind—just like solving an insight problem!

Henry was incapable of that type of creativity and insight. He uttered no *aha*s and constructed alternative meanings for virtually none of the sentences in my ambiguity-detection experiment. Henry's overuse of clichés in that 1966 study also illustrated the demise of his *expressive* creativity. Both findings raised a question that engaged my thinking and research off and on for the next fifty years: does the hippocampus support the everyday creativity seen in linguistic comprehension and expression? *Remembering* summarizes my answer to that question.

To even discuss everyday creativity, I told Samantha, I had to lobby for an expanded definition that counts any kind of new and useful idea as creative. New to *the brain that created the idea* but not necessarily new to every human brain ever. And useful to *someone*, but not necessarily to humankind at large. That enlarged concept of creativity challenges the Romantic-era framework that dominated thinking about creativity for over two hundred years—a framework that restricted the term *creative* to genius-level ideas that are novel in the history of the world, thereby eliminating the possibility that everyday creativity might occur.

Dismantling that framework helped me publish my research on Henry's inability to comprehend what made captioned cartoons funny—findings that contradicted the familiar anecdotes about Henry's keen sense of humor. Consistent with my experimental results, none of the anecdotes held up. All allowed simpler, unwitty interpretations. This paved the way to examining Henry's ability to comprehend metaphors resembling Shakespeare's "Juliet is the sun"—a genuine product of *genius-level* creativity. Unlike control participants, Henry could not understand such metaphors—a finding that opened the door to examining Henry's ability to create new and useful ideas in ordinary life. His deficits in that final experiment completed the picture: the normal creativity inherent in every aspect of human life—including comprehension—reflects engagement of the hippocampus.

BIG THEORY, BROWSERS, BINDING MECHANISMS, AND THE BRAIN

Remembering also develops a Big Theory that makes sense of issues mentioned only briefly in my 2014 interview with Samantha. One is the question of how the normal brain retrieves already formed memories for words. Everyday speech errors of normal folks unlocked that mystery. An unintended word that intrudes as a speech error virtually always belongs to the same category as the word the speaker intended. Someone wanting to utter the word *battle* (a *common* noun) might inadvertently substitute the common noun *bottle* in error, but not *Betty* (a *proper* noun). Because errors in all languages examined so far obey that like-for-like category rule, this showed that brain browsers—the activating mechanisms in the frontal lobes for quickly retrieving words—are designed to activate specific classes of words such as common nouns or proper nouns.

This idea allows insights into aphasia. People with damage to the frontal brain region that houses browsers can experience difficulty retrieving entire *categories* of words and phrases. Damage to your *pronoun* browser can render you unable to retrieve pronouns, but perfectly capable of retrieving common nouns. Damage to your *common noun* browser can render you unable to use common nouns, but still able to use pronouns. Unlike category-specific aphasics, Henry had no difficulty retrieving high-frequency words in any category. All of the browsers in his frontal lobe were intact.

The Big Theory in *Remembering* also explains how the normal brain forms new memories. Unlocking that mystery required two keys. One was a clear distinction between what counts as new versus old. An *old* stimulus has an internal representation in the perceiver's brain that is *preformed* and *functional* for *accomplishing the task at hand*. A *novel* stimulus lacks a functional internal representation for *performing the current task*. A penny is a new stimulus when you try to draw all of its features from memory, but it represents an old stimulus in financial transactions—the typical task involving pennies.

Henry provided the second key to unlocking how the hippocampus creates new memories. His errors show that binding mechanisms in the hippocampus normally form new concepts by conjoining two or more preformed concepts in particular categories. This binding function of the hippocampus is what allows ordinary folks to create novel internal representations for grammatical sentences, for personally experienced events, and for organized visual objects such as faces.

Henry's errors also established that binding mechanisms are designed to combine *particular categories of concepts*. To show this, *Remembering* summarized some surprising new discoveries about the binding mechanism that creates internal representations for proper names consisting of a first, middle, and last name. Henry learned proper names such as *Lee Harvey Oswald* that he could only have encountered after his operation. He also linked novel information to those proper names—just like a memory-normal person.[7] That could only happen if Henry's operation spared his hippocampal binding mechanisms for linking proper names to associated information, and if binding mechanisms in the hippocampus create memories for specific categories of information.

CELEBRATION, COMMEMORATION, AND HELPING YOURSELF

Besides deepening and extending ideas in my 2014 interview with Samantha, *Remembering* adds some new dimensions: gratitude, celebration, commemoration, and helping others. *Remembering* commemorates the scientific revolution whereby Henry forever reshaped the intellectual landscape of memory, mind, and brain, and helped everyone understand what memory is, how it works, and its role in everything that makes your human mind and brain worth owning. *Remembering* also celebrates Henry's contribution to the emerging focus on the mind and brain of *individuals* such as himself rather than on the "average" human mind and brain—a mythical object.[8] Finally, *Remembering* highlights the research with Henry that shows how you, the reader, can enhance and maintain your memory, mind, and brain in ways that Henry could not. Let's conclude by examining those ways.

WHAT YOU CAN DO THAT HENRY COULD NOT

- create new and useful internal representations in comprehension and action;
- relearn and restore information in your brain degraded by aging and disuse;
- remember brief events you experienced seconds, minutes, or hours earlier;

- quickly form new internal representations to register and correct your own and others' errors;
- quickly detect unexpected events and novelty in general;
- read, understand, and learn new words at age seventy-five and beyond;
- create a program of physical exercise to keep sharp the millions of memories stored in your brain;
- participate in intellectual discussions;
- initiate conversations to stay socially connected;
- access the Internet to help you retrieve forgotten information;
- generate novel sentences that are relevant, coherent, accurate, grammatical, and comprehensible;
- read and define familiar but uncommon words as an older adult;
- name pictures of familiar but rarely encountered objects, again as an older adult;
- learn new skills such as aqua-aerobics and writing for a general audience;
- participate in social organizations such as an author's club;
- volunteer to help students learn English;
- create an effective learning plan that repeats newly encountered information at gradually increasing intervals spaced across minutes, hours, or days;
- recall the irregular spelling of familiar but rare words at age seventy or older;
- forget information at a normal rather than faster-than normal rate;
- ask questions calling for additional information from a conversational partner;
- ask questions to help you link a newly encountered person to her name;
- detect unfamiliar target forms in complex visual arrays;
- comprehend and generate metaphors that connect new concepts to familiar ones;
- understand who did what to whom in sentences;
- analyze your errors so as not to repeat them;

- comprehend and deliberately generate apt and well-timed humor;
- select and organize complex information to create new internal representations that are distinctive, concise, and memorable.

THE KEEP-ACTIVE APPROACH TO MAINTAINING MEMORY, MIND, AND BRAIN

Two easy-to-remember words summarize how you can maintain your memory, mind, and brain and avoid the rapid declines with aging that Henry experienced: *Keep Active*—physically, socially, and intellectually. Keep developing your skills. Keep learning. Even when unable to recall, actively try to remember.

Nowhere does *Remembering* list the thousands of hands-on ways you can apply the *Keep Active* rule in your everyday physical, social, and intellectual activities. What would be the point? You would soon forget that list. Unlike computers, human brains are not good at remembering lists. Even if you could upload *my* list directly into your brain, would you retrieve it? Only if you organized and generated the facts *yourself* and used them day in and day out like a common word would you remember them over an extended period.

Already scattered throughout *Remembering* are the physical, social, and intellectual activities that *I* enjoy for enhancing and maintaining *my* memories. However, *your* memories and the everyday physical, social, and intellectual activities *you* enjoy probably differ from mine. Besides, self-generated information is better remembered than knowledge passively received, as you may recall. The *Keep Active* approach to maintaining memories is intended to help you learn to think, create, evaluate, organize, apply, and use information—skills that can last a lifetime. Unlike fact-learning, normal individuals never completely forget well-learned skills!

So, at the heart of *Remembering* lies a dilemma. *Remembering* can only meaningfully fulfill the promise that knowledge about Henry will help others if you the reader actively help yourself. Can we escape that dilemma? Yes, if you put to full use the sections labeled *Test Your Memory* and *Ques-*

tions for Reflection that end each chapter in *Remembering.* If you skipped some of those questions when reading the book, I urge you to revisit the ones marked **. They are not about remembering facts in the chapter you just read. They are opportunities for you to enhance your memory, mind, and brain by actively applying the *Keep Active* rule on your own. To illustrate, this was the first of nine double-asterisk questions in Chapter 2: *What enjoyable activities besides solving crossword puzzles would you recommend for maintaining language memories as people grow older?***

The caveat *besides solving crossword puzzles* is intended to remind you of research discussed in Chapter 2: everyone loves solving puzzles, but no evidence indicates that *crossword* puzzles will improve your ability to remember.[9] Unravelling clues such as *a word referring to the capitol of California, ten letters long, with A in position two and M in position six* will not greatly enhance your indelible memory for how to spell *Sacramento.* Learned as a child and used ever since, that knowledge is unforgettable.

Next comes the action-oriented question itself: *What enjoyable activities would you recommend for maintaining language memories?* Like the nearly one hundred double-asterisk questions in *Remembering*, answering that one will engage your ability to think about, imagine, and elaborate on information in deep ways that will likely help you maintain and enhance your memories over an extended period. And since you *enjoy* those activities, you are likely to incorporate them into your everyday life! By helping yourself in this way you will help everyone who promised that research with Henry will help others (see Reflection Box 26.1).

REFLECTION BOX 26.1: WARNING SIGNS INDICATING THAT YOUR HIPPOCAMPUS IS HEALTHY

1) You delight in *acquiring new information and new skills.*
2) You want to *meet new people* and experience fresh points of view.
3) You *appreciate humor* and generate appropriate and well-timed wit.

4) You tend to *adapt creatively to changing circumstances.*
5) You strive to *create new ideas*, especially ones that are helpful to others.
6) You quickly *repair errors in perception and action*, your own and others'.
7) You are skilled at *avoiding isolation and staying socially connected.*
8) You love *metaphors that link new concepts with familiar ones.*
9) You readily *organize novel information* into distinct, concise and memorable images.
10) You *create flexible life goals* that are appropriate to your age, skills and passions.

TEST YOUR MEMORY FOR CHAPTER 26

1) Can you recall the subtitle of *Remembering* word-for-word? (Cover page.)

QUESTIONS FOR REFLECTION: CHAPTER 26

1) If you did not recall the subtitle of *Remembering* word-for-word, why not?**
2) If you did remember the subtitle of *Remembering*, how did you do this?**

ACKNOWLEDGMENTS

Remembering would not exist without three helpful people: Hans-Lukas Teuber, Suzanne Corkin and Henry Molaison. I thank Professor Teuber for introducing me to Henry in 1966. I thank Dr. Corkin for inviting my lab to test Henry at the MIT Clinical Research Center in the late 1990s. And I especially thank Henry for participating tirelessly, and unbeknownst to me, without remuneration, in my experiments.

Endless thanks and love goes to my wife, Deborah May Burke, for encouraging and supporting every aspect of this research and book-writing project, from beginning to end. Without her help, I could not have written *Remembering*. Virtually every chapter benefited from Deborah's wisdom and extensive knowledge of the scientific literature on memory and cognitive aging. By providing thoughtful comments on multiple drafts of the book, sometimes at extremely short notice, Deborah also helped me meet what seemed like impossible publisher deadlines.

Another huge set of thanks comes to my friends in the Claremont Writing Workshop, especially Don Coleman, Judith Favor, Daniella Franco, Connie Green, Lissa Petersen, Peggy Redman, John Rogers, Laura St. Martin, Hart St. Martin, Barbara Smythe and Teri Tomkins. Their constructive critiques of my early drafts taught me how to write non-academic prose. I also thank Judith Favor and Lissa Petersen for the compassion and sagacity of their observations and suggestions after reading *Remembering* in its entirety. Much appreciated too was the helpful thoughts, advice, comments, or encouragement from Alan Castel, John Berry, Gus Craik, Luke Dittrich, Jim Geiwitz, Lori James, Barbara Jay, Cynthia Kenyon, Samantha Kimmey, Richard Kirschman, Herb Kutchins, Ivan Light, Laura Lopez, Doris Ober, Alex Riley, Jasper Rine, Meredith Shafto, Lolly Tyler, William Marslen-Wilson, and my son, Kennen MacKay.

I also thank William Marslen-Wilson for providing paper copies of the transcript of his 1970 interviews with Henry and for allowing Lori James and I to post this important information on the Internet. Thanks likewise to Luke Dittrich for providing the useful photographs he took of Henry's 1982 responses to questionnaires in the Beck Depression Inventory. Special appreciation too goes to Lissa Petersen and Jim Geiwitz. Their honest appraisal of the *Reflection Boxes* and *Illustration Boxes* in the advanced reader copy of *Remembering* helped me redesign and revise those sidebars to engage a wider audience. Thanks as well to members of my target audience, too numerous to mention by name, who served as an international focus group for evaluating draft aspects of *Remembering*, including its proposed titles and cover designs.

However, it is to my outstanding research assistant, Micah Alan Johnson, that I wish to express my foremost gratitude. To encourage readers to reflect on and apply what they learned to their personal life, Micah drafted the *Questions for Reflection* at the end of each chapter. To help readers consolidate what they learned, Micah drafted the *Test your Memory Questions* that ended the chapters. Micah also conducted analyses of Henry's 1982 responses on the Beck Depression Inventory—statistics essential for settling a scientific dispute about Henry's everyday state of mind in the years following his operation. By researching and summarizing the many recent findings on humor, laughter, and the brain, Micah also greatly strengthened the chapter on Henry's "keen sense of humor." Micah even assisted with the automation of citations, copyediting, content editing, and proofreading the book, constructing novel figures, compiling chapters into a book file, obtaining high resolution images and reprint permissions, and recruiting a graduate student artist to create additional figures from scratch. For all this and more, I thank Micah.

Next to last, I thank Cynthia Zigmund, literary agent with Second City Publishing Services in Chicago. Cynthia negotiated the book contract for *Remembering* and helped me navigate the sometimes treacherous straits and choppy waters of the highly competitive world of publishing. Lastly, I apologize to friends whose help with this project over the past fifty years I have forgotten. By reminding me, your name will appear in the second edition of *Remembering*.

Appendix

TIMELINE OF MY RESEARCH WITH HENRY AND MAJOR EVENTS IN HIS LIFE

February 1926: Henry's date of birth.

1947: Henry graduates from high school.

August 1953: Henry consents to and undergoes brain surgery in Hartford, Connecticut.

Spring 1965: As a twenty-three year old MIT graduate student, I meet Henry and test his ability to detect and explain the two meanings of ambiguous sentences.

Spring 1972: I distribute to friends and colleagues at UCLA an unpublished manuscript comparing comprehension, memory and attention in Henry versus the thirty Harvard undergraduates who participated in my ambiguity detection experiment.

December 1980: Henry enters the Bickford Health Care Center—his home for the rest of his life.

December 1981: Henry's mother dies.

1992–2015: Based on Henry's errors in comprehending and producing sentences in my 1966 study, I design and conduct a series of experiments in collaboration with Dr. Deborah Burke, WM Keck Professor of Psychology and chair of the Linguistics and Cognitive Science Department at Pomona College, UCLA postdoctoral fellows Jennifer Taylor and Elizabeth Graham, UCLA graduate students Meredith Shafto, Lise Abrams, and Laura Johnson, and undergraduates Marat Ahmetzanov, Joel Schwartz, Diane Marian, and Jennifer Dyer. Given Henry's amygdala damage, our goal was to understand the role of the amygdala and emotion in perceiving and recalling taboo versus neutral words presented at extremely rapid rates. The basic conclusion of our studies was that emotional information receives priority over neutral information in the process of forming new memories. In short, when time is limited, the amygdala sets priorities for the hippocampus.

1995–2013: My UCLA lab, the lab of Dr. Lori James at the University of Colorado, Colorado Springs, and the lab of Dr. Deborah Burke recruit about sixty individuals with age-normal memory who resembled Henry in age, education, socio-economic status, background and intelligence. From 1995 until 2013, these control participants performed twenty-five tasks that closely resembled the ones that Henry completed at MIT from 1966 to 1999.

Winter 1997: I design and Dr. Lori James runs the first seven of the twenty-five MIT experiments that tested, among other things, Henry's ability to define uncommon words, to distinguish words from non-words, and to spell and correctly pronounce familiar words that were irregularly spelled.

1998–2010: Based on Henry's sentence comprehension errors in my 1966 study, I design and conduct a series of experiments in collaboration with UCLA graduate student Meredith Shafto. The goal of our experiments is to understand why normal individuals miscomprehend questions resembling *How many animals of each kind did Moses take on the ark?* The basic conclusion of our studies is that accurate comprehension of novel information is a creative process that requires engagement of the hippocampus, unlike routine comprehension involving existing internal representations for words such as Moses and Noah.

Winter 1998: I design and Dr. Lori James runs six experiments to test, among other things, Henry's ability to describe pictures on the Test of Language Competence, and to read aloud ambiguous and unambiguous sentences and high versus low frequency words presented in isolation.

1998: With coauthors Deborah Burke and Rachel Stewart, a graduate student at Stanford University, I publish our research findings on Henry's ability to produce sentences that are novel, accurate and coherent in explaining the meanings of ambiguous sentences.

1998: Coauthors Rachael Stewart, Deborah Burke and I publish our research findings on Henry's ability to comprehend the two meanings of ambiguous sentences.

Winter 1999: I design and Dr. Lori James runs twelve experiments to test, among other things, Henry's ability to describe captioned cartoons and to name pictures on the Boston Naming Test.

2001: Coauthor Dr. Lori James and I publish our experiments showing how Henry's word knowledge declined with aging at a faster than normal rate.

2002: Coauthor Dr. Lori James and I publish our experimental findings on the unusual effects of aging on Henry's memory for the phonology and spelling of words.

2006: I publish a review article on how aging impacted Henry's memory and language.

2007: With coauthors Dr. Lori James, UCLA post-doctoral fellow Jenifer Taylor, and Diane Marion, a graduate student at the University of California Berkeley, I publish our research findings indicating that Henry's language and memory exhibit parallel deficits and sparing.

2008: Coauthors Dr. Lori James, UCLA graduate student Chris Hadley, and I publish our initial research findings on Henry's language performance on the Test of Language Competence.

December 2008: Henry dies of congestive heart failure at Bickford Health Care Center.

2009: Coauthor Chris Hadley and I publish our research findings on how Henry's retrograde amnesia for familiar word memories worsened with aging.

2009: Coauthor Lori James and I publish our experiments on Henry's visual cognition deficits in the what's-wrong-here and Hidden-Figures tasks.

2011: With coauthors Dr. Lori James, graduate students Chris Hadley and Katherine Fogler, I publish our analyses of Henry unusual speech errors in the Test of Language Competence and in describing what makes captioned cartoons funny.

2013: Coauthors Dr. Lori James, graduate students Laura Johnson and

Vedad Fazel, and I publish our analyses of how Henry compensated for some of his sentence production deficits on the Test of Language Competence.

2013: Graduate student coauthors Laura Johnson and Chris Hadley and I publish our research findings on Henry's surprising ability to remember proper names and other categories of information.

2013: Coauthor Laura Johnson and I publish our theoretical review article on Henry's inability to detect and correct speech errors, his own and others'.

2016: With coauthor Rudy Goldstein, a graduate student at Kansas University, I publish data comparing the linguistic creativity of Henry versus other celebrities participating in real world interviews.

NOTES

Chapter 1: We're All Forgetters, Samantha

1. A verbatim transcript of my interview with Samantha Kimmey appears in Samantha Kimmey, "Don MacKay: A Pioneer at the Intersection of Language and Memory," *Point Reyes Light*, no. 67 (May 1, 2014): 1–11. Although the gist of Samantha's questions are as quoted, I have expanded some of my answers for this book.

2. D. G. MacKay, "The Engine of Memory: Even after His Death, the Famous Amnesic H.M. Is Revolutionizing Our Understanding of How Memory Works and How We Maintain It as We Age," *Scientific American Mind* 25, no. 3 (2014): 30–38.

3. Luke Dittrich, *Patient H.M.: A Story of Memory, Madness, and Family Secrets* (New York: Random House, 2016).

4. Sigmund Freud, *The Psychopathology of Everyday Life* (Berlin, 1901).

5. The received wisdom has since changed as these recent publications citing my research with Henry illustrate:

V. Piai, K. L. Anderson, J. J. Lin, et al., "Direct Brain Recordings Reveal Hippocampal Rhythm Underpinnings of Language Processing," *Proceedings of the National Academy of Sciences* 113, no. 40 (2016): 11366–71; P. Meyer, A. Mecklinger, T. Grunwald, et al., "Language Processing within the Human Medial Temporal Lobe," *Hippocampus* 15, no. 4 (2005): 451–59; N. V. Covington and M. C. Duff, "Expanding the Language Network: Direct Contributions from the Hippocampus," *Trends in Cognitive Sciences* 20, no. 12 (2016): 869–70.

6. D. G. MacKay, "The Earthquake That Reshaped the Intellectual Landscape of Memory, Mind and Brain: Case H.M.," in *Cases of Amnesia: Contributions to Understanding Memory and the Brain*, ed. S. E. MacPherson and S. Della Sala (New York: Taylor and Francis, forthcoming).

7. V. W. Sunghasette, M.C. Friedman, and A. Castel, "Memory and Metamemory for Inverted Words: Illusions of Competency and Desirable Difficulties Memory and Metamemory for Inverted Words: Illusions of Competency and Desirable Difficulties," *Psychonomic Bulletin Review* 18, no. 5 (2011): 973–78.

Chapter 2: Weird News, Inadequately Researched

1. Benedict Carey, "No Memory, but He Filled in the Blanks," *New York Times*, December 6, 2010.

2. This book adopts standard conventions for representing the spelling and pronunciation

of words and syllables. Capitalization indicates a word as spelled, e.g., BICYCLE, or as depicted, e.g., a picture of a BICYCLE. Capitalization *within* a word indicates primary syllabic stress and dots indicate syllable boundaries. For example, PEDESTRIAN is pronounced "pe•DES•tree•an."

3. See D. G. MacKay, "The Engine of Memory: Even after His Death, the Famous Amnesic H.M. Is Revolutionizing Our Understanding of How Memory Works and How We Maintain It as We Age," *Scientific American Mind* 25, no. 3 (2014): 30–38.

4. W. Marslen-Wilson, "Biographical Interviews with H.M." (unpublished transcript; MIT, Cambridge, MA, 1970), http://mackay.bol.ucla.edu/. Posted with permission from William Marslen-Wilson.

5. G. A. Carlesimo and M. Oscar-Berman, "Memory Deficits in Alzheimer's Patients: A Comprehensive Review," *Neuropsychology Review* 3 (1992): 119–69.

6. E. R. Kandel, *In Search of Memory: The Emergence of a New Science of Mind* (New York: Norton, 2006), pp. 332–35.

7. *Diagnostic and Statistical Manual of Mental Disorders*, 5th ed. (DSM–5) (Philadelphia, PA: American Psychiatric Association, 2013); S. Salmans, *Depression: Questions You Have; Answers You Need. People's Medical Society* (Allentown, PA: Rodale Press, 1997).

8. L. E. James and D. G. MacKay, "H.M., Word Knowledge, and Aging: Support for a New Theory of Long-Term Retrograde Amnesia," *Psychological Science* 12 (2001): 485–92; N. Hunkin et al., "Focal Retrograde Amnesia Following Closed Head Injury: A Case Study and Theoretical Account," *Neuropsychologia* 33, no. 4 (1995): 509–23; M. Kopelman, "Focal Retrograde Amnesia and the Attribution of Causality: An Exceptionally Critical Review," *Cognitive Neuropsychology* 17, no. 7 (2000): 585–621; E. Tulving et al., "Priming of Semantic Autobiographical Knowledge: A Case Study of Retrograde Amnesia," *Brain and Cognition* 8, no. 1 (1988): 3–20; N. Kapur et al., "Focal Retrograde Amnesia Following Bilateral Temporal Lobe Pathology," *Brain* 115, no. 1 (1992): 73–85.

9. S. Corkin, *Permanent Present Tense: The Unforgettable Life of the Amnesic Patient, H.M.* (New York: Basic Books, 2013), p. xx.

10. Word frequency is the technical term for how often a word occurs within some large and specifiable corpus of written text, say, an extensive sample of books, magazines, or newspaper articles containing thirty million words written for adults. For example, *triage*, *thimble*, and *pretzel* are low-frequency words because they occur less than once per million in a very large database of Associated Press articles; see B. R. Postle and S. Corkin, "Impaired Word-Stem Completion Priming but Intact Perceptual Identification Priming with Novel Words: Evidence from the Amnesic Patient H.M.," *Neuropsychologia* 36 (1998): 421–40.

11. L. E. James and D. G. MacKay, "H.M., Word Knowledge and Aging: Support for a New Theory of Long-Term Retrograde Amnesia," *Psychological Science* 12 (2001): 485–92.

12. Ibid.

13. Ibid.

14. D. G. MacKay and L. E. James, "Aging, Retrograde Amnesia, and the Binding Problem for Phonology and Orthography: A Longitudinal Study of 'Hippocampal Amnesic' H.M.," *Aging, Neuropsychology, and Cognition* 9 (2002): 298–333.

15. D. G. MacKay and C. Hadley, "Supra-Normal Age-Linked Retrograde Amnesia: Lessons from an Older Amnesic (H.M.)," *Hippocampus* 19 (2009): 424–45.

16. Ibid.

17. Ibid.

18. Ibid.; D. G. MacKay and L. E. James, "The Binding Problem for Syntax, Semantics, and Prosody: H.M.'s Selective Sentence-Reading Deficits under the Theoretical-Syndrome Approach," *Language and Cognitive Processes* 16 (2001): 419–60; MacKay and James, "Aging, Retrograde Amnesia."

19. Carey, "No Memory, but He Filled in the Blanks."

20. B. G. Skotko, D. C. Rubin, and L. A. Tupler, "H.M.'s Personal Crossword Puzzles: Understanding Memory and Language," *Memory* 16, no. 2 (2008): 89–96.

21. Dr. Suzanne Corkin's emailed refusal to share H.M.'s crossword puzzles was dated December 19, 2005, 5:49 p.m. She gave this reason for refusing: the puzzles "contain information that would reveal confidential information about H.M. We are dedicated to protecting his privacy above all else." The emailed information about procedures for disciplining violators of the data-sharing guidelines of the American Psychological Association (APA) came from the Chief APA Editorial Advisor, dated December 29, 2005, 11:06 a.m.

22. Henry's books included *Pocket Patter of Dell Puzzle*, the *Penny Press Means Puzzle Pleasure*, *Large Print Crosswords* and other books for beginners: See Skotko, Rubin, and Tupler, "H.M.'s Personal Crossword Puzzles."

23. B. G. Skotko et al., "Puzzling Thoughts for H.M.: Can New Semantic Information Be Anchored to Old Semantic Memories?" *Neuropsychology* 18 (2004): 756–69.

24. Ibid.

25. Unlike myself, others have expressed a desire to know more. Did the *New York Times* make the story up? Did the *New York Times* interviewees simply fabricate? Did the 2008 research report fabricate? Do Henry's crossword books really exist? Only this last question has an answer. A reliable witness saw some of Henry's puzzle books at the MIT Clinical Research Center in 1997 and in a personal communication characterized them as "fifth grade level."

26. S. Corkin, *Permanent Present Tense: The Unforgettable Life of the Amnesic Patient, H.M.* (New York: Basic Books, 2013), p. xx.

27. See for example, M. Gazzaniga, R. Ivry, and G. Mangun, *Cognitive Neuroscience: The Biology of the Mind* (New York: W. W. Norton, 2009).

28. Corkin, *Permanent Present Tense.*

29. L. Dittrich, *Patient H.M.: A Story of Memory, Madness, and Family Secrets* (New York: Random House, 2016). The recording of Dr. Corkin speaking about shredding can be heard at lukedittrich.com.

Chapter 3: How Can You Help Vulnerable Memories Survive?

1. L. Ossher, K. E. Flegal, and C. Lustig, "Everyday Memory Errors in Older Adults," *Aging, Neuropsychology, and Cognition* 20 (2013): 220–42.

2. L. E. James and D. M. Burke, "Phonological Priming Effects on Word Retrieval and Tip-of-the-Tongue Experiences in Young and Older Adults," *Journal of Experimental Psychology: Learning, Memory, and Cognition* 26 (2000): 1378–91.

3. D. Z. Hambrick, T. A. Salthouse, and E. J. Meinz, "Predictors of Crossword Puzzle Proficiency and Moderators of Age-Cognition Relations," *Journal of Experimental Psychology: General* 128 (1999): 131–64; D. G. MacKay and L. Abrams, "Age-Linked Declines in Retrieving Orthographic Knowledge: Empirical, Practical, and Theoretical Implications," *Psychology and Aging* 13 (1998): 647–62; J. W. Rowe and R. L. Kahn, *Successful Aging* (New York: Pantheon, 1998).

4. Mary Margaret Groves, *Facia: An Artwork* (Claremont, CA: Pomona College Senior Art Exhibit, 2016).

5. Atul Gawande, "The Heroism of Incremental Care," *New Yorker*, January 23, 2017.

6. Ibid.

7. For a short overview of relevant research, see C. Chabris and D. Simons, *The Invisible Gorilla: And Other Ways Our Intuitions Deceive Us* (New York: Crown, 2010).

Chapter 4: Creative Aging: A Silver Lining

1. For elaborations on this theme, see C. Honore, *In Praise of Slowness: Challenging the Cult of Speed* (New York: HarperCollins, 2004).

2. D. M. Burke et al., "On the Tip of the Tongue: What Causes Word Finding Failures in Young and Older Adults?" *Journal of Memory and Language* 30 (1991): 542–79.

3. Deborah M. Burke compared TOT resolution for two groups who had experienced a TOT in response to a question such as "What is the word that refers to a type of interlocking fastener made of nylon": an experimental group that had the question repeated on several subsequent trials and then tried to retrieve the target word, versus a control group that did not get question repeated after the initial TOT response. A week later the experimental group was reliably more likely than the control group to answer the question correctly, even when participants had not retrieved the target word earlier. See also J. M. Gardiner, F. I. M. Craik, and F. A. Bleasdale, "Retrieval Difficulty and Subsequent Recall," *Memory & Cognition* 1 (1973): 213–16; H. L. Roediger III and J. D. Karpicke, "The Power of Testing Memory: Basic Research and Implications for Educational Practice," *Perspectives on Psychological Science* 1, no. 3 (2006): 181–210.

4. D. G. MacKay, L. Abrams, and M. J. Pedroza, "Aging on the Input Versus Output Side: Age-Linked Asymmetries between Detecting versus Retrieving Orthographic Information," *Psychology and Aging* 14 (1999): 3–17; see also D. G. MacKay and L. Abrams, "Age-Linked Declines in Retrieving Orthographic Knowledge: Empirical, Practical, and Theoretical Implications," *Psychology and Aging* 13 (1998): 647–62.

5. D. G. MacKay and C. Hadley, "Supra-Normal Age-Linked Retrograde Amnesia: Lessons from an Older Amnesic (H.M.)," *Hippocampus* 19 (2009): 424–45; D. G. MacKay and L. E. James, "The Binding Problem for Syntax, Semantics, and Prosody: H.M.'s Selective Sentence-Reading Deficits under the Theoretical-Syndrome Approach," *Language and Cognitive Processes* 16 (2001): 419–60; D. G. MacKay and L. E. James, "Aging, Retrograde Amnesia, and the Binding Problem for Phonology and Orthography: A Longitudinal Study of 'Hippocampal Amnesic' H.M.," *Aging, Neuropsychology, and Cognition* 9 (2002): 298–333; L. E. James and

D. G. MacKay, "H.M., Word Knowledge and Aging: Support for a New Theory of Long-Term Retrograde Amnesia," *Psychological Science* 12 (2001): 485–92.

6. J. D. E. Gabrieli, N. J. Cohen, and S. Corkin, "The Impaired Learning of Semantic Knowledge Following Bilateral Medial Temporal-Lobe Resection," *Brain and Cognition* 7 (1988): 157–77.

7. MacKay and Hadley, "Supra-Normal Age-Linked Retrograde Amnesia."

8. S. Corkin, "Lasting Consequences of Bilateral Medial Temporal Lobectomy: Clinical Course and Experimental Findings in H.M.," *Seminars in Neurology* 4 (1984): 249–59. Because this study did not compare H.M.'s BNT performance with that of matched controls, the basis for this "without difficulty" claim is unknown.

9. E. A. Kensinger, M. T. Ullman, and S. Corkin, "Bilateral Medial Temporal Lobe Damage Does Not Affect Lexical or Grammatical Processing: Evidence from Amnesic Patient H.M.," *Hippocampus* 11 (2001): 347–60. Because this study compared H.M.'s BNT performance to his own previous BNT performance rather than with that of matched controls, it did not (in principle) demonstrate accelerating deterioration with aging (see Appendix A).

10. MacKay and Hadley, "Supra-Normal Age-Linked Retrograde Amnesia."

11. B. R. Postle and S. Corkin, "Impaired Word-Stem Completion Priming but Intact Perceptual Identification Priming with Novel Words: Evidence from the Amnesic Patient H.M.," *Neuropsychologia* 36 (1998): 421–40.

12. MacKay and Hadley, "Supra-Normal Age-Linked Retrograde Amnesia."

13. As a statistical concept in neuropsychology, standard deviation refers to the variation in test performance of members of a control group relative to their expected performance, namely the mean for the group as a whole. Thus, a difference of one standard deviation or less relative to a control group indicates that H.M.'s performance was close to the mean or typical performance of normal individuals, and a difference of two standard deviations indicates that his performance was extreme relative to typical members of the control group. By standard convention, performance differences are considered reliable or statistically significant when a patient's test scores differ from the mean score for control individuals by two or more standard deviations.

14. W. Marslen-Wilson and H. L. Teuber, "Memory for Remote Events in Anterograde Amnesia: Recognition of Public Figures from News Photographs," *Neuropsychologia* 13 (1975): 353–64.

15. Corkin, "Lasting Consequences of Bilateral Medial Temporal Lobectomy."

16. L. L. Carstensen et al., "Emotional Experience in Everyday Life across the Adult Life Span," *Journal of Personal and Social Psychology* 79 (2000): 644–55.

17. A. Gawande, "The Heroism of Incremental Care," *New Yorker*, January 23, 2017.

18. D. J. Simons et al., "Do 'Brain-Training' Programs Work?" *Psychological Science in the Public Interest* 17, no. 3 (2016).

19. L. Fratiglioni, "Influence of Social Network on Occurrence of Dementia: A Community-Based Longitudinal Study," *Lancet* 355 (2000), pp. 1315–19.

20. M. Gonzalles-Gross, A. Marcos, and K. Pietrzik, "Nutrition and Cognitive Impairment in the Elderly," *British Journal of Nutrition* 86, no. 3 (2001), pp. 313–21; and M. L. Correa Leite et al., "Nutrition and Cognitive Deficit in the Elderly: A Population Study," *European Journal of Clinical Nutrition* 55, no. 12 (2001), pp. 1053–58.

21. Yaffe et al., "A Prospective Study of Physical Activity and Cognitive Decline in Elderly Women: Women Who Walk," *Archives of Internal Medicine* 161, no. 14, pp. 1703–708. See also D. Lauren et al., "Physical Activity and Risk of Cognitive Impairment and Dementia in Elderly Persons," *Archives of Neurology* 58, no. 3 (2001), pp. 498–504.

22. Marquis et al., "Independent Predictors of Cognitive Decline in Healthy Elderly Persons," *Archives of Neurology* 59, no. 4 (2002), pp. 601–606.

23. B. Fischhoff and R. Beyth, "'I Knew It Would Happen': Remembered Probabilities of Once-Future Things," *Organizational Behavior and Human Performance* 13 (1975), pp. 1–16;

See also A. D. Castel, et al., "The Dark Side of Expertise: Domain-Specific Memory Errors," *Psychological Science* 18, no. 1 (2007), pp. 3–8.

24. To further test this theory of memory maintenance, my lab hopes to determine whether other amnesics with hippocampal damage also experience exaggerated or faster-than-normal degradation of their memories for rarely used information as they age. We also want to find out whether healthy older adults, after reencountering forgotten information in natural everyday settings, routinely rejuvenate memory components that aging and disuse have destroyed.

25. Three years is the average lag between onset of amnesia and neuropsychological tests.

Chapter 5: What Is It Like to Be You, Henry?

1. L. Dittrich, *Patient H.M.: A Story of Memory, Madness, and Family Secrets*, 2nd ed. (New York: Random House, 2017).

2. L. E. James and D.G. MacKay, "H.M., Word Knowledge, and Aging: Support for a New Theory of Long-Term Retrograde Amnesia," *Psychological Science* 12 (2001): 485–92.

3. D. G. MacKay et al., "Amnesic H.M. Exhibits Parallel Deficits and Sparing in Language and Memory: Systems versus Binding Theory Accounts," *Language and Cognitive Processes* 22, no. 3 (2007): 377–452.

4. Dittrich, *Patient H.M.*

5. Ibid.

6. W. Marslen-Wilson, "Biographical Interviews with H.M." (unpublished transcript; MIT, Cambridge, MA, 1970). Posted at http://mackay.bol.ucla.edu/ with permission from William Marslen-Wilson.

7. P. J. Hilts, *Memory's Ghost: The Strange Tale of Mr. M. and the Nature of Memory* (New York: Simon & Schuster, 1995).

8. Dittrich, *Patient H.M.*

9. R. Adolphs et al., "Impaired Recognition of Emotion in Facial Expressions Following Bilateral Damage to the Human Amygdala," *Nature* 15 (1994): 669–72.

10. S. Corkin, *Permanent Present Tense: The Unforgettable Life of the Amnesic Patient, H.M.* (New York: Basic Books, 2013), p. 111.

11. James J. DiCarlo, letter to the editor of the *New York Times* magazine, August 5, 2016.

12. B. G. Skotko et al., "Puzzling Thoughts for H.M.," *Neuropsychology* 18 (2004): 756–69. See also Corkin, *Permanent Present Tense*.

13. Dr. DiCarlo had written that Professor Corkin worked in the final days before her death to organize and preserve all of H.M.'s MIT records, contrary to what Dittrich wrote about shredding in *Patient H. M.*

14. Dittrich, *Patient H.M.*

Chapter 6: Welcome to Henry's Lifesaving Contract with Science!

1. S. Corkin, *Permanent Present Tense: The Unforgettable Life of the Amnesic Patient, H.M.* (New York: Basic Books, 2013).

2. Ibid.

3. E. J. Bourne. *The Anxiety and Phobia Workbook*, 5th ed. (Oakland, CA: New Harbinger Publications, 2010).

Chapter 7: The New Memory Factory

1. Michael Lewis, *The Undoing Project: The Friendship That Changed Our Minds* (New York: W. W. Norton, 2017).

2. D. G. MacKay, D. M. Burke, and R. Stewart, "H.M.'s Language Production Deficits: Implications for Relations between Memory, Semantic Binding, and the Hippocampal System," *Journal of Memory and Language* 38 (1998): 28–69; D. G. MacKay, R. Stewart, and D. M. Burke, "H.M. Revisited: Relations between Language Comprehension, Memory, and the Hippocampal System," *Journal of Cognitive Neuroscience* 10 (1998): 377–94; D. G. MacKay and L. E. James, "Visual Cognition in Amnesic H.M.: Selective Deficits on the What's-Wrong-Here and Hidden-Figure Tasks," *Journal of Experimental and Clinical Neuropsychology* 31 (2009): 769–89.

3. Lewis, *Undoing Project*.

4. Ibid.

5. See for example, S. Corkin, *Permanent Present Tense: The Unforgettable Life of the Amnesic Patient, H.M.* (New York: Basic Books, 2013).

6. D. G. MacKay, "Stage Theories Refuted," in *A Companion to Cognitive Science*, ed. W. Bechtel and G. Graham (Oxford: Blackwell, 1998), pp. 671–78.

7. G. Rizzolatti, "The Mirror-Neuron System and Imitation," in *Perspectives on Imitation: From Mirror Neurons to Memes*, ed. S. Hurley and N. Chater (Cambridge, MA: MIT Press, 2004); G. Rizzolatti et al., "Premotor Cortex and the Recognition of Motor Actions," *Cognitive Brain Research* 3, no. 131 (1996): 131–41.

8. D. G. MacKay, "The Problem of Rehearsal or Mental Practice," *Journal of Motor Behavior* 13, no. 4 (1981): 274–85. See also D. G. MacKay, *The Organization of Perception and Action: A Theory for Language and Other Cognitive Skills* (Berlin: Springer-Verlag, 1987), pp. 1–254. Also available at http://mackay.bol.ucla.edu/publications.html.

9. MacKay, Burke, and Stewart, "H.M.'s Language Production Deficits"; MacKay, Stewart, and Burke, "H.M. Revisited."

10. S. E. MacPherson and S. Della Sala, eds., *Cases of Amnesia: Contributions to Understanding Memory and the Brain* (New York: Taylor and Francis, forthcoming).

11. X. Liu et al., "Optogenetic Stimulation of a Hippocampal Engram Activates Fear Memory Recall," *Nature* 484 (2012): 381–85.

12. MacKay, *Organization of Perception and Action*.

13. This penny memory demonstration originated with Raymond A. Nickerson and M. J. Adams, "Long Term memory for a Common Object," *Cognitive Psychology* 11 (1979): 287–307.

14. B. Milner, "Memory and the Temporal Regions of the Brain," in *Biology of Memory*, ed. K. H. Pribram and D. E. Broadbent (New York: Academic Press, 1970).

Chapter 8: Fetch that Memory, Browser

1. R. Meringer and K. Mayer, *Versprechen und Verlesen: Eine Psychologisch-Linguistische Studie* (Stuttgart: Goschensche Verlagsbuchhandlung, 1895); see also R. Meringer, *Aus dem Leben der Sprache: Versprechen; Kindersprache, Nachahmungstrieb* (Berlin: Behr's Verlag, 1908).

2. D. M. Burke and M. A. Shafto, "Language and Aging," in *The Handbook of Aging and Cognition*, ed. F. I. M. Craik and T. A. Salthouse (New York: Psychology Press, 2004), pp. 373–443.

3. E. H. Sturtevant, *Linguistic Science* (New Haven, CT: Yale University Press, 1947).

4. Daniel L. Schacter, *The Seven Sins of Memory: How the Mind Forgets and Remembers* (New York: Houghton Mifflin, 2001).

5. Victoria Fromkin, ed., *Errors of Linguistic Performance: Slips of the Tongue, Ear, Pen, and Hand* (New York: Academic Press, 1980), pp. 319–32; Victoria Fromkin, ed., *Speech Errors* (Amsterdam: Mouton Press, 1973).

6. See Meringer, *Aus dem Leben der Sprache*.

7. E. Loftus, *Eyewitness Testimony* (Cambridge, MA: Harvard University Press, 1979).

8. R. L. Marsh, J. D. Landau, and J. L. Hicks, "Contributions of Inadequate Source Monitoring to Unconscious Plagiarism during Idea Generation," *Journal of Experimental Psychology: Learning, Memory, and Cognition* 23 (1997): 886–97.

9. Fromkin, ed., *Errors of Linguistic Performance*; Fromkin, ed., *Speech Errors*.

10. Steven Pinker, *The Sense of Style* (New York: Viking, 2014), p. 12.

11. Steven Pinker, *How the Mind Works* (New York: Norton, 1997).

12. Ibid.

13. D. G. MacKay, "Spoonerisms: The Structure of Errors in the Serial Order of Speech," *Neuropsychologia* 8 (1970): 323–50.

14. D. G. MacKay et al., "Compensating for Language Deficits in Amnesia I: H.M.'s Spared Retrieval Categories," *Brain Sciences* 3, no. 1 (2013): 262–93; D. G. MacKay, *The Organization of Perception and Action: A Theory for Language and Other Cognitive Skills* (New York: Springer-Verlag, 1987), pp. 44–61.

15. Fromkin, ed., *Errors of Linguistic Performance*.

16. Meringer and Mayer, *Versprechen und Verlesen*; see also Meringer, *Aus dem Leben der Sprache*.

17. H. Goodglass and E. Kaplan, *The Assessment of Aphasia and Related Disorders* (Baltimore, MD: Williams & Wilkins, 1983), p. 76.

18. Ibid.

19. A. Caramazza et al., "The Organization of Lexical Knowledge in the Brain: Evidence from Category- and Modality-Specific Deficits," in *Mapping the Mind*, ed. L. A. Hirschfeld and S. A. Gelman (Cambridge, UK: Cambridge University Press, 1994), pp. 68–84; R. Job, M. Miozzo, and G. Sartori, "On the Existence of Category-Specific Impairments. A Reply to Parkin and Stewart," *Quarterly Journal of Experimental Psychology Section A* 46 (1993): 511–16.

20. Goodglass and Kaplan, *Assessment of Aphasia and Related Disorders*.

21. D. G. MacKay and L. E. James, "Aging, Retrograde Amnesia, and the Binding Problem for Phonology and Orthography: A Longitudinal Study of 'Hippocampal Amnesic' H.M.," *Aging, Neuropsychology, and Cognition* 9 (2002): 298–333; D. G. MacKay and L. E. James, "The Binding Problem for Syntax, Semantics, and Prosody: H.M.'s Selective Sentence-Reading Deficits under the Theoretical-Syndrome Approach," *Language and Cognitive Processes* 16 (2001): 419–60; D. G. MacKay, L. E. James, and C. B. Hadley, "Amnesic H.M.'s Performance on the Language Competence Test: Parallel Deficits in Memory and Sentence Production," *Journal of Clinical and Experimental Neuropsychology* 30 (2008): 280–300; D. G. MacKay, J. Taylor, and D. E. Marian, "Amnesic H.M. Exhibits Parallel Deficits and Sparing in Language and Memory: Systems versus Binding Theory Accounts," *Language and Cognitive Processes* 22 (2007): 377–452; D. G. MacKay, D. M. Burke, and R. Stewart, "H.M.'s Language Production Deficits: Implications for Relations between Memory, Semantic Binding, and the Hippocampal System," *Journal of Memory and Language* 38 (1998): 28–69; D. G. MacKay, *The Organization of Perception and Action: A Theory for Language and Other Cognitive Skills* (New York: Springer-Verlag, 1987), pp. 44–61.

22. Daniel Kahneman, *Thinking, Fast and Slow* (New York: Farrar, Straus & Giroux, 2011), pp. 1–499.

23. D. M. Burke and D. G. MacKay, "Memory, Language, and Ageing," *Philosophical Transactions of the Royal Society of London: Biological Sciences* 352 (1997): 1845–56; R. T. Krampe and K. A. Ericsson, "Maintaining Excellence: Deliberate Practice and Elite Performance in Young and Older Pianists," *Journal of Experimental Psychology: General* 125 (1996): 331–59; N. Charness et al., "The Role of Deliberate Practice in Chess Expertise," *Applied Cognitive Psychology* 19 (2005): 151–65.

24. S. Higgs, A. C. Williamson, and A. S. Attwood, "Recall of Recent Lunch and Its Effect on Subsequent Snack Intake," *Physiology & Behavior* 94 (2008): 454–62; S. Higgs and J. E. Donohue, "Focusing on Food during Lunch Enhances Lunch Memory and Decreases Later Snack Intake," *Appetite* 57 (2011): 202–206; L. R. Vartanian, W. H. Chen, N. M. Reily, and A. D. Castel, "The Parallel Impact of Episodic Memory and Episodic Future Thinking on Food Intake," *Appetite* 101 (2016): 31–36.

25. A. D. Castel, "Think More, Eat Less? Memory Can Make Us Eat Less," *Metacognition and the Mind* (blog), *Psychology Today*, May 9, 2016; Vartanian, Chen, Reily, and Castel, "Parallel Impact of Episodic Memory."

26. D. Z. Hambrick, T. A. Salthouse, and E. J. Meinz, "Predictors of Crossword Puzzle Proficiency and Moderators of Age-Cognition Relations," *Journal of Experimental Psychology* 12, no. 2 (1999): 131–64.

27. See, for example, K. Ball et al., "Long-Term Effects of Cognitive Training on Everyday Functional Outcomes in Older Adults," *Journal of the American Medical Association* 296, no. 23 (2006): 2805–14; C. Chabris and D. Simons, *The Invisible Gorilla: And Other Ways Our Intuitions Deceive Us* (New York: Crown, 2010); S. Colcombe and A. F. Kramer, "Fitness Effects on the Cognitive Function of Older Adults," *Psychological Science* 14 (2003): 125–40; S. Colcombe et al., "Aerobic Exercise Training Increases Brain Volume in Aging Humans," *Journal of Gerontology: Medical Sciences* 61 (2006): 1166–70; C. Hertzog et al., "Enrichment Effects on Adult Cognitive Development," *Psychological Science in the Public Interest* 9 (2009): 1–65; A. F. Kramer et al., "Aging, Fitness, and Neurocognitive Function," *Nature* 400 (July 29, 1999): 418–19; A. F. Kramer and K. I. Erickson, "Capitalizing on Cortical Plasticity: Influence of Physical Activity on Cognition and Brain Function," *Trends in Cognitive Sciences* 11, no. 8 (2007): 342–48; S. L. Willis, "Cognitive Training and Everyday Competence," in *Annual Review of Gerontology and Geriatrics*, ed. K. W. Schaie, vol. 7 (New York: Springer, 1987); F. D. Wolinsky et al., "The ACTIVE Cognitive Training Trial and Health-Related Quality of Life: Protection That Lasts for 5 Years," *Journals of Gerontology: Series A, Biological Sciences and Medical Sciences* 61 (2006): 1324–29.

28. R. S. Wilson et al., "Participating in Stimulating Cognitive Activities and Risk of Incident Alzheimer's Disease," *Journal of the American Medical Association* 287, no. 6 (2002): 742–48; A. Merluzzi, "Cognitive Shields: Investigating Protections against Dementia," *Observer, Association for Psychological Science* 28, no. 2 (2015): 21–24.

29. E. J. Bourne, *The Anxiety and Phobia Workbook*, 5th ed. (Oakland, CA: New Harbinger Publications, 2010); A. D. Castel, "10,000 Simple Steps to a Better Memory: Take a Walk; Walking Can Help Memory, and Can Have Many Other Health Benefits," *Metacognition and the Mind* (blog), *Psychology Today*, January 9, 2014.

30. S. Corkin, *Permanent Present Tense: The Unforgettable Life of the Amnesic Patient, H.M.* (New York: Basic Books, 2013).

Chapter 9: Humpty Dumpty after His Fall

1. Plato, *Theaetetus*, trans. R. A. Waterford (Harmondsworth, 1987).

2. As quoted in M. D. Lemonick, *The Perpetual Now: A Story of Amnesia, Memory, and Love* (New York: Doubleday, 2017).

3. S. Corkin, *Permanent Present Tense: The Unforgettable Life of the Amnesic Patient, H.M.* (New York: Basic Books, 2013).

4. Ibid.

5. M. G. Packard and B. J. Knowlton, "Learning and Memory Functions of the Basal Ganglia," *Annual Review in Neuroscience* 25 (2002): 563–93.

6. Lemonick, *Perpetual Now*.

7. Corkin, *Permanent Present Tense*.

8. B. Milner, "Memory Impairment Accompanying Bilateral Hippocampal Lesions," in *Psychologie De L'hippocampe*, ed. P. Passouant (Paris, France: Centre National de la Recherche Scientifique, 1962), pp. 257–72.

9. D. G. MacKay and L. E. James, "Aging, Retrograde Amnesia, and the Binding Problem for Phonology and Orthography: A Longitudinal Study of 'Hippocampal Amnesic' H.M.," *Aging, Neuropsychology, and Cognition* 9 (2002): 298–333.

10. C. J. Steele and V. B. Penhune, "Specific Increases within Global Decreases: A Functional Magnetic Resonance Imaging Investigation of Five Days of Motor Sequence Learning," *Journal of Neuroscience* 30, no. 24 (2010): 8332–41; see also G. Albouy et al., "Both the Hippocampus and Striatum Are Involved in Consolidation of Motor Sequence Memory," *Neuron* 58 (2008): 261–72; R. Poldrack, "Sequence Learning: What's the Hippocampus to Do?" *Neuron* 37, no. 6 (2003): 891–93; H. E. Schendan et al., "An fMRI Study of the Role of the Medial Temporal Lobe in Implicit and Explicit Sequence Learning," *Neuron* 37, no. 6 (2003): 1013–25.

11. D. G. MacKay and C. Hadley, "Supra-Normal Age-Linked Retrograde Amnesia: Lessons from an Older Amnesic (H.M.)," *Hippocampus* 19 (2009): 424–45; D. G. MacKay and L. W. Johnson, "Errors, Error Detection, Error Correction and Hippocampal-Region Damage: Data and Theories," *Neuropsychologia* (2013): 2633–50.

Chapter 10: Putting Humpty Dumpty Together Again

1. For more details on these procedures, see WikiHow, s.v. "How to Ride a Bicycle," last updated September 22, 2018, http://www.wikihow.com/Ride-a-Bicycle (accessed September 4, 2017).

2. M. A. Lynch, "Long-Term Potentiation and Memory," *Physiological Reviews* 84, no. 1 (2004): 87–136.

3. Steven Pinker, *The Stuff of Thought: Language As a Window into Human Nature* (New York: Viking, 1998), pp. 241–42.

4. C. J. Steele and V. B. Penhune, "Specific Increases within Global Decreases: A Functional Magnetic Resonance Imaging Investigation of Five Days of Motor Sequence Learning," *Journal of Neuroscience* 30, no. 24 (2010): 8332–41.

5. D. G. MacKay and L. E. James, "Visual Cognition in Amnesic H.M.: Selective Deficits on the What's-Wrong-Here and Hidden-Figure Tasks," *Journal of Experimental and Clinical Neuropsychology* 31 (2009): 769–89; D. G. MacKay, L. E. James, and C. B. Hadley, "Amnesic H.M.'s Performance on the Language Competence Test: Parallel Deficits in Memory and Sentence Production," *Journal of Clinical and Experimental Neuropsychology* 30 (2008): 280–300; D. G. MacKay, "The Problem of Flexibility, Fluency, and Speed-Accuracy Trade-Off in Skilled Behavior," *Psychological Review* 89, no. 5 (1982): 483–506.

6. Sourcewatch, s.v. "Repetition," last edited December 18, 2007, http://www.source watch.org/index.php?title=Repetition.

7. Ibid.

8. D. G. MacKay, G. Wulf, C. Yin, and L. Abrams, "Relations between Word Perception and Production: New Theory and Data on the Verbal Transformation Effect," *Journal of Memory and Language* 32 (1993): 624–46.

SECTION III: AWARENESS OUT OF THE BLUE

1. Billy Collins, "The Blue," in *The Apple That Astonished Paris* (Fayetteville: University of Arkansas Press, 1988).

Chapter 11: Can You Create New Concepts, Henry?

1. R. K. Sawyer, *Explaining Creativity: The Science of Human Innovation* (New York: Oxford University Press, 2012).
2. D. G. MacKay and T. G. Bever, "In Search of Ambiguity," *Perception and Psychophysics* 2 (1967): 193–200.
3. D. G. MacKay, R. Stewart, and D. M. Burke, "H.M. Revisited: Relations between Language Comprehension, Memory, and the Hippocampal System," *Journal of Cognitive Neuroscience* 10 (1998): 377–94.
4. See D. G. MacKay, "The Search for Ambiguity by an Amnesic Patient: Implications for the Theory of Comprehension, Memory, and Attention" (unpublished manuscript; Psychology Department, University of California, Los Angeles, 1972). See also MacKay, Stewart, and Burke, "H.M. Revisited." Also available at http://mackay.bol.ucla.edu/publications.html.
5. Ibid.
6. D. G. MacKay, L. E. James, J. K. Taylor, and D. E. Marian, "Amnesic H.M. Exhibits Parallel Deficits and Sparing in Language and Memory: Systems versus Binding Theory Accounts," *Language and Cognitive Processes* 22, no. 3 (2007): 377–452.
7. Ibid.
8. L. A. Stowe, A. M. J. Paans, A. A. Wijers, and F. Zwarts, "Activations of 'Motor' and Other Non-Language Structures during Sentence Comprehension," *Brain and Language* 89 (2004): 290–99.
9. R. B. Ivry and S. W. Keele, "Timing Functions of the Cerebellum," *Journal of Cognitive Neuroscience* 1 (1989): 136–52.
10. D. W. Zaidel et al., "The Interpretation of Sentence Ambiguity in Patients with Unilateral Focal Brain Surgery," *Brain Language* 51 (1995): 458–68.
11. D. G. MacKay and R. Goldstein, "Creativity, Comprehension, Conversation and the Hippocampal Region: New Data and Theory," *AIMS Neuroscience* 3, no. 1 (2016): 105–42, DOI: 10.3934/Neuroscience.2016.1.105. Available at http://www.aimspress.com/journal/neuroscience.

Chapter 12: Creative Comprehension in Everyday Life

1. D. G. MacKay and R. Goldstein, "Creativity, Comprehension, Conversation and the Hippocampal Region: New Data and Theory," *AIMS Neuroscience* 3, no. 1 (2016): 105–42, DOI: 10.3934/Neuroscience.2016.1.105. Available at http://www.aimspress.com/journal/neuroscience.
2. Ibid.

3. Of course the ability to create the all-important context of "Juliet is the sun" requires, in addition, the genetics, passion, and dedication of a Shakespeare. See also S. Pinker, *How the Mind Works* (New York: W. W. Norton, 1997).

4. Julie Sedivy, *Language in Mind: An Introduction to Psycholinguistics* (Irvine, CA: Oxford University Press, 2017).

5. MacKay and Goldstein, "Creativity, Comprehension, Conversation and the Hippocampal Region."

Chapter 13: The Man Who Mistook a Wastebasket for a Window

1. Oliver Sacks, *The Man Who Mistook His Wife for a Hat* (New York: Touchstone, 1998).

2. B. Milner, S. Corkin, and H. L. Teuber, "Further Analysis of the Hippocampal Amnesic Syndrome: 14-Year Follow-Up Study of H.M.," *Neuropsychologia* 6 (1968): 215–34.

3. L. L. Thurstone, *A Factorial Study of Perception* (Chicago: University of Chicago Press, 1949). The original source for Figure 13.1 appears in T. D. Ben-Soussan, A. Berkovich-Ohana, J. Glicksohn, and A. Goldstein, "A Suspended Act: Increased Reflectivity and Gender-Dependent Electrophysiological Change Following Quadrato Motor Training," *Frontiers in Psychology*, 5 (2014), https://www.frontiersin.org/articles/10.3389/fpsyg.2014.00055/full#B36. Its original source is the database used by J. Glicksohn, and Z. Kinberg, "Performance on Embedded Figure Tests: Profiling Individual Differences," *Journal of Individual Differences* 30, no. 3 (2009): 152–62, https://insights.ovid.com/individual-differences/jindif/2009/30/030/performance-embedded-figures-tests/5/01222908.

4. Milner, Corkin, and Teuber, "Further Analysis of the Hippocampal Amnesic Syndrome."

5. D. G. MacKay and L. E. James, "Visual Cognition in Amnesic H.M.: Selective Deficits on the What's-Wrong-Here and Hidden-Figure Tasks," *Journal of Experimental and Clinical Neuropsychology* 31 (2009): 769–89.

6. T. Tallarico, *The Haunted House: What's Wrong Here?* (Newburyport, MA: Kidsbooks, 1991).

7. MacKay and James, "Visual Cognition in Amnesic H.M."

8. D. G. MacKay, L. E. James, C. B. Hadley, and K. A. Fogler, "Speech Errors of Amnesic H.M.: Unlike Everyday Slips-of-the-Tongue," *Cortex* 47 (2011): 377–408.

9. D. G. MacKay, J. Taylor, and D. E. Marian, "Amnesic H.M. Exhibits Parallel Deficits and Sparing in Language and Memory: Systems versus Binding Theory Accounts," *Language and Cognitive Processes* 22 (2007): 377–452.

10. MacKay and James, "Visual Cognition in Amnesic H.M."

11. Milner, Corkin, and Teuber, "Further Analysis of the Hippocampal Amnesic Syndrome."

12. MacKay and James, "Visual Cognition in Amnesic H.M."

13. Ibid.

14. Ibid.
15. Ibid.
16. Ibid.

Chapter 14: The Mysterious Face in the Mirror

1. Oliver Sacks, "Why Are Some of Us Terrible at Recognizing Faces?" *New Yorker*, August 30, 2010.
2. Ibid.
3. Ibid.
4. Ibid.
5. Ibid.
6. A. Ishai, C. F. Schmidt, and P. Boesiger, "Face Perception Is Mediated by a Distributed Cortical Network," *Brain Research Bulletin* 67, no. 1–2 (2005): 87–93; J. V. Haxby, E. A. Hoffman, and M. I. Gobbini, "The Distributed Human Neural System for Face Perception," *Trends in Cognitive Science* 4, no. 6 (2000): 223–33; J. M. DeGutis, S. Bentin, L. C. Robertson, and M. D'Esposito, "Functional Plasticity in Ventral Temporal Cortex Following Cognitive Rehabilitation of a Congenital Prosopagnosic," *Journal of Cognitive Neuroscience* 19, no. 11 (2007): 1790–1802.
7. Oliver Sacks, *The Man Who Mistook His Wife for a Hat* (New York: Touchstone, 1998).
8. D. G. MacKay, L. E. James, J. Taylor, and D. E. Marian, "Amnesic H.M. Exhibits Parallel Deficits and Sparing in Language and Memory: Systems versus Binding Theory Accounts," *Language and Cognitive Processes* 22 (2007): 377–452.
9. Sacks, "Why Are Some of Us Terrible at Recognizing Faces?"
10. This review appeared in Salon magazine, with an excerpt on the first un-numbered page of Oliver Sacks, *On the Move* (New York: Vintage, 2015).
11. A. Treisman and G. Gelade, "A Feature-Integration Theory of Attention," *Cognitive Psychology* 12 (1980): 97–136.
12. Ibid.
13. D. G. MacKay and L. E. James, "Visual Cognition in Amnesic H.M.: Selective Deficits on the What's-Wrong-Here and Hidden-Figure Tasks," *Journal of Experimental and Clinical Neuropsychology* 31 (2009): 769–89.
14. A. Treisman, "Features and Objects: The Fourteenth Bartlett Memorial Lecture," *Quarterly Journal of Experimental Psychology* 40A (1988): 201–36.
15. Ibid.
16. M. Shafto and D. G. MacKay, "The Moses, Mega-Moses, and Armstrong Illusions: Integrating Language Comprehension and Semantic Memory," *Psychological Science* 11 (2000): 372–78; M. Shafto and D. G. MacKay, "Miscomprehension, Meaning, and Phonology: The Unknown and Phonological Armstrong Illusions," *European Journal of Cognitive Psychology* 22, no. 4 (2010): 529–68.
17. Ibid.
18. Ibid.

19. P. S. Eriksson, E. Perfilieva, T. Bjork-Eriksson, A.-M. Alborn, C. Nordborg, D. A. Peterson, and F. H. Gage, "Neurogenesis in the Adult Human Hippocampus," *Nature Medicine* 4, no. 11 (1998), pp. 1313–17, https://www.societyns.org/runn/2009/pdfs/bednarsept2109neurogenesis1998.pdf;

C. V. Dennis, L. S. Suh, M. L. Rodriguez, J. J. Kril, and G. T. Sutherland, "Human Adult Neurogenesis Across the Ages: An Immunohistological Study," *Neuropathology and Applied Neurobiology* 42, no. 7 (2016), pp. 621–38, https://onlinelibrary.wiley.com/doi/full/10.1111/nan.12337; K. L. Spalding et al., "Dynamics of Hippocampal Neurogenesis in Adult Humans," *Cell* 153, no. 6 (2013), pp. 1183–84. https://www.sciencedirect.com/science/article/pii/S0092867413005333;

J. S. Biane and M. A. Kheirbek, "Imaging Adult Hippocampal Neurogenesis in Vivo," *Neuropsychopharmacology* 42, no. 373 (2017), https://www.nature.com/articles/npp2016200;

L. K. Smith, C. W. White, and S. A. Villeda, "The Systematic Environment: At the Interface of Aging and Adult Neurogenesis," *Cell and Tissue Research* 371, no. 1 (2018): 105–13, https://link.springer.com/article/10.1007/s00441-017-2715-8; but see S. F. Sorrells et al., "Human Hippocampal Neurogenesis Drops Sharply in Children to Undetectable Levels in Adults," *Nature* 555 (2018): 377–81, https://www.nature.com/articles/nature25975.

Chapter 15: Metaphors Be with You, Henry

1. G. Lakoff and M. Johnson, *Metaphors We Live By* (New York: W. W. Norton, 1980).
2. S. Pinker, *How the Mind Works* (New York: W. W. Norton, 1997).
3. Ibid.
4. T. Konishi, "The Semantics of Grammatical Gender: A Cross-Cultural Study," *Journal of Psycholinguistic Research* 22 (1993): 519–34; T. Konishi, "The Connotations of Gender: A Semantic Differential Study of German and Spanish," *Word* 45 (1994): 317–27; D. G. MacKay and T. Konishi, "Personification and the Pronoun Problem," *Women's Studies International Quarterly* 3 (1980): 149–63; D. G. MacKay and T. Konishi, "Contraconscious Internal Theories Influence Lexical Choice during Sentence Completion," *Consciousness and Cognition* 3 (1994): 196–222; D. G. MacKay and T. Konishi, "The Selection of Pronouns in Spoken Language Production: An Illusion of Reference," in *Reflecting Sense: Perception and Appearances in Literature, Culture, and the Arts*, ed. F. Burwick and W. Pape (Berlin: Walter & De Gruyter, 1995), pp. 279–300; also available at http://mackay.bol.ucla.edu/publications.html.
5. Konishi, "Semantics of Grammatical Gender"; Konishi, "Connotations of Gender."
6. Ibid.; MacKay and Konishi, "Personification and the Pronoun Problem"; MacKay and Konishi, "Contraconscious Internal Theories"; MacKay and Konishi, "Selection of Pronouns"; see also D. G. MacKay, "Gender in English, German, and Other Languages: Problems with the Old Theory, Opportunities for the New," in *Perceiving and Performing Gender: Wahrnehmung und Herstellung von Geschlecht*, ed. U. Pasero and F. Braun (Wiesbaden: VS Verlag für Sozialwissenschaften, 1999), pp. 73–87. Also available at http://www.mackay.bol.ucla.edu/publications; D. G. MacKay, "Language, Thought, and Social Attitudes," in *Language: Social Psychological Perspectives*, ed. H. Giles, W. P. Robinson, and P. M. Smith (Oxford, England: Pergamon, 1980), pp. 89–96; D. G. MacKay, "On the

Goals, Principles, and Procedures for Prescriptive Grammar," *Language in Society* 9 (1980): 349–67; D. G. MacKay, "Protypicality among Metaphors: On the Relative Frequency of Personification and Spatial Metaphors in Literature Written for Children versus Adults," *Metaphor and Symbolic Activity* 1, no. 2 (1986): 87–107.

7. E. H. M. Wiig and W. Secord, *Test of Language Competence: The Metaphor Subtest of the Expanded Edition* (New York: The Psychological Corporation, Harcourt, Brace, Jovanovich, 1988).

8. D. G. MacKay, L. E. James, and C. Hadley, "Amnesic H.M.'s Performance on the Language Competence Test: Parallel Deficits in Memory and Sentence Production," *Journal of Experimental and Clinical Neuropsychology* 30 (2008): 280–300.

9. Ibid.

10. W. Deng, J. B. Aimone, and F. H. Gage, "New Neurons and New Memories: How Does Adult Hippocampal Neurogenesis Affect Learning and Memory?" *Nature Review Neuroscience* 11, no. 5 (2010): 339–50.

11. S. Pinker, *How the Mind Works* (New York: W. W. Norton, 1997).

Chapter 16: What's in a Name, Henry?

1. J. R. Lackner, "Observations on the Speech Processing Capabilities of an Amnesic Patient: Several Aspects of H.M.'s Language Function," *Neuropsychologia* 12 (1974): 199–207.

2. D. G. MacKay, "To End Ambiguous Sentences," *Perception and Psychophysics* 1 (1966): 426–36; see D. G. MacKay, "The Search for Ambiguity by an Amnesic Patient: Implications for the Theory of Comprehension, Memory, and Attention" (unpublished manuscript; Psychology Department, University of California Los Angeles, 1972). Also available at http://mackay.bol.ucla.edu/publications.html.

3. Lackner, "Observations on the Speech Processing Capabilities."

4. D. G. MacKay, R. Stewart, and D. M. Burke, "H.M. Revisited: Relations between Language Comprehension, Memory, and the Hippocampal System," *Journal of Cognitive Neuroscience* 10, no. 3 (1998): 377–94; D. G. MacKay, D. M. Burke, and R. Stewart, "H.M.'s Language Production Deficits: Implications for Relations between Memory, Semantic Binding, and the Hippocampal System," *Journal of Memory and Language* 38 (1998): 28–69.

5. Ibid.

6. Luke Dittrich, *Patient H.M.: A Story of Memory, Madness, and Family Secrets* (New York: Random House, 2016).

7. S. Corkin, *Permanent Present Tense: The Unforgettable Life of the Amnesic Patient, H.M.* (New York: Basic Books, 2013).

8. That claim also seemed dubious because published results indicated that the patient's brain damage extended all the way from the hippocampal region to the frontal lobes and his performance on tests of frontal lobe function was severely impaired. See D. G. MacKay, "A Tale of Two Paradigms or Metatheoretical Approaches to Cognitive Neuropsychology. Did Schmolck, Stefanacci, and Squire Demonstrate That 'Detection and Explanation of Sentence Ambiguity Are Unaffected by Hippocampal Lesions but Are Impaired by Larger Temporal Lobe Lesions'?" *Brain and Language* 78 (2001): 265–72.

9. Stuart A. Kirk and Herb Kutchins, *The Selling of DSM* (New York: Walter de Gruyter, 1992).

10. American Psychiatric Association, *Diagnostic and Statistical Manual of Mental Disorders: DSM-III-TR* (Washington, DC: American Psychiatric Association, 1994).

For a review of the growing body of evidence for a genetic basis to homosexuality, see Dean Hamer and Peter Copeland, *The Science of Desire: The Search for the Gay Gene and the Biology of Behavior* (New York: Simon & Schuster, 1995).

11. Kirk and Kutchins, *Selling of DSM.*

12. Daniel Barron, "Why Psychiatry Needs Neuroscience," *Scientific American* (blog), April 25, 2017, https://blogs.scientificamerican.com/guest-blog/why-psychiatry-needs-neuroscience/.

13. Steven Pinker, *How the Mind Works* (New York: Norton, 1997).

14. Kirk and Kutchins, *Selling of DSM.*

15. Pinker, *How the Mind Works.*

16. Ibid.

17. D. G. MacKay. "The Earthquake That Reshaped the Intellectual Landscape of Memory, Mind and Brain: Case H.M.," in *Cases of Amnesia: Contributions to Understanding Memory and the Brain*, ed. S. E. MacPherson and S. Della Sala (New York: Taylor and Francis, forthcoming).

Chapter 17: The Hippocampus Has a Shadow, Henry

1. P. S. Churchland and T. J. Sejnowski, *The Computational Brain* (Cambridge, MA: MIT Press, 1992); Steven Pinker, *How the Mind Works* (New York: Norton, 1997).

2. Pinker, *How the Mind Works.*

3. See D. G. MacKay, R. Stewart, and D. M. Burke, "H.M. Revisited: Relations between Language Comprehension, Memory, and the Hippocampal System," *Journal of Cognitive Neuroscience* 10, no. 3 (1998): 377–94; D. G. MacKay, D. M. Burke, and R. Stewart, "H.M.'s Language Production Deficits: Implications for Relations between Memory, Semantic Binding, and the Hippocampal System," *Journal of Memory and Language* 38 (1998): 28–69.

4. D. G. MacKay and L. E. James, "The Binding Problem for Syntax, Semantics, and Prosody: H.M.'s Selective Sentence-Reading Deficits under the Theoretical-Syndrome Approach," *Language and Cognitive Processes* 16 (2001): 419–60; D. G. MacKay and L. E. James, "Aging, Retrograde Amnesia, and the Binding Problem for Phonology and Orthography: A Longitudinal Study of 'Hippocampal Amnesic' H.M.," *Aging, Neuropsychology, and Cognition* 9 (2002): 298–333; D. G. MacKay, J. Taylor, and D. E. Marian, "Amnesic H.M. Exhibits Parallel Deficits and Sparing in Language and Memory: Systems versus Binding Theory Accounts," *Language and Cognitive Processes* 22 (2007): 377–452; D. G. MacKay and C. Hadley, "Supra-Normal Age-Linked Retrograde Amnesia: Lessons from an Older Amnesic (H.M.)," *Hippocampus* 19 (2009): 424–45.

5. C. J. Steele and V. B. Penhune, "Specific Increases Within Global Decreases: A Functional Magnetic Resonance Imaging Investigation of Five Days of Motor Sequence Learning,"

Journal of Neuroscience 30, no. 24 (2010): 8332–41; G. Albouy, V. Sterpenich, E. Balteau, et al., "Both the Hippocampus and Striatum Are Involved in Consolidation of Motor Sequence Memory," *Neuron* 58 (April 24, 2008): 261–72; R. Poldrack, "Sequence Learning: What's the Hippocampus to Do?" *Neuron* 37, no. 6 (2003): 891–93; H. E. Schendan, M. M. Searl, R. J. Melrose, and C. E. Stern, "An fMRI Study of the Role of the Medial Temporal Lobe in Implicit and Explicit Sequence Learning," *Neuron* 37, no. 6 (2003): 1013–25; see also chapter 7.

6. D. G. MacKay, "Perception, Action, and Awareness: A Three-Body Problem," in *Relationships between Perception and Action*, ed. O. Neumann and W. Prinz (Berlin: Springer-Verlag, 1990), pp. 269–303.

7. Pinker, *How the Mind Works*.

8. Only "uncommitted neurons" in the cortex can represent *novel* conjunctions of information such as *Sarah + Macpherson* for someone unfamiliar with that word combination. However, committed neurons are "committed" to representing a familiar piece of information in the cortex such as the isolated word *Sarah*. Henry's inability to activate uncommitted neurons for prolonged periods implies that his consciousness differed from normal, but not that he was unconscious. This is because the hippocampus achieves prolonged activation and awareness of novel information indirectly, via inhibition that allows committed neurons to prolong the "reverberated" activation that is inevitable with hierarchically organized mirror neurons. Such prolonged, reverberated activation of committed neurons allowed Henry to become aware of the familiar but not the novel information that he encountered after his operation. See MacKay, "Perception, Action, and Awareness," also D. G. MacKay, D. M. Burke, and R. Stewart, "H.M.'s Language Production Deficits: Implications for Relations Between Memory, Semantic Binding, and the Hippocampal System," *Journal of Memory and Language* 38 (1998): 28–69; D. G. MacKay, R. Stewart, and D. M. Burke, "H.M. Revisited: Relations between Language Comprehension, Memory, and the Hippocampal System," *Journal of Cognitive Neuroscience* 10 (1998): 377–94. Also available at http://mackay.bol.ucla.edu/publications.html.

Chapter 18: She Said What to Whom, Henry?

1. Gary Larson, "Raising the Dead," *The Far Side*, date unknown; D. G. MacKay, L. E. James, C. B. Hadley, and K. A. Fogler, "Speech Errors of Amnesic H.M.: Unlike Everyday Slips-of-the-Tongue," *Cortex* 47 (2011): 377–408.

2. D. G. MacKay, "The Search for Ambiguity by an Amnesic Patient: Implications for the Theory of Comprehension, Memory, and Attention" (unpublished manuscript; Psychology Department, University of California, Los Angeles, 1972). Available at http://mackay.bol.ucla.edu/publications.html.

3. E. H. M. Wiig and W. Secord, "The 'Thematic Roles' Subtest," *Test of Language Competence* (New York: The Psychological Corporation, Harcourt, Brace, Jovanovich, 1988). Also see D. G. MacKay, L. E. James, and C. Hadley, "Amnesic H.M.'s Performance on the Language Competence Test: Parallel Deficits in Memory and Sentence Production," *Journal of Experimental and Clinical Neuropsychology* 30 (2008): 280–300.

4. D. G. MacKay, L. E. James, J. Taylor, and D. E. Marian, "Amnesic H.M. Exhibits

Parallel Deficits and Sparing in Language and Memory: Systems versus Binding Theory Accounts," *Language and Cognitive Processes* 22 (2007): 377–452.

5. J. R. Lackner, "Observations on the Speech Processing Capabilities of an Amnesic Patient: Several Aspects of H.M.'s Language Function," *Neuropsychologia* 12 (1974): 199–207.

6. Ibid.

7. See D. G. MacKay, R. Stewart, and D. M. Burke, "H.M. Revisited: Relations between Language Comprehension, Memory, and the Hippocampal System," *Journal of Cognitive Neuroscience* 10, no. 3 (1998): 377–94.

8. Ibid.

9. Ibid.

10. MacKay, James, Taylor, and Marian, "Amnesic H.M. Exhibits Parallel Deficits."

11. Ibid. For follow-up findings with other amnesics and contrasting procedures that support this "new internal representation" interpretation of hippocampal processing, see the following:

M. D. Barense, D. Gaffan, and K. S. Graham, "The Human Medial Temporal Lobe Processes Online Representations of Complex Objects," *Neuropsychologia* 45 (2007): 2963–74; D. Hannula, D. Tranel, and N. J. Cohen, "The Long and the Short of It: Relational Memory Impairments in Amnesia, Even at Short Lags," *Journal of Neuroscience* 26 (2006): 8352–59; I. R. Olson, K. S. Moore, M. Stark, and A. Chatterjee, "Visual Working Memory Is Impaired When the Medial Temporal Lobe Is Damaged," *Journal of Cognitive Neuroscience* 18 (2006): 1087–97; C. Ranganath and M. D'Esposito, "Medial Temporal Lobe Activity Associated with Active Maintenance of Novel Information," *Neuron* 31 (2001): 865–73; D. E. Warren, M. C. Duff, D. Tranel, and N. J. Cohen, "Medial Temporal Lobe Damage Impairs Representation of Simple Stimuli," *Frontiers in Human Neuroscience* 4, no. 35 (May 18, 2010): 1–9.

12. MacKay, James, Taylor, and Marian, "Amnesic H.M. Exhibits Parallel Deficits."

13. For reviews relevant to this point, see the following:

D. G. MacKay, "The Earthquake That Reshaped the Intellectual Landscape of Memory, Mind and Brain: Case H.M.," in *Cases of Amnesia: Contributions to Understanding Memory and the Brain*, ed. S. E. MacPherson and S. Della Sala (New York: Taylor and Francis, forthcoming); D. G. MacKay and R. Goldstein, "Creativity, Comprehension, Conversation and the Hippocampal Region: New Data and Theory," *AIMS Neuroscience* 3, no. 1 (2016): 105–42. Available at http://www.aimspress.com/journal/neuroscience.

14. Steven Pinker, *The Language Instinct* (New York: William Morrow, 1994).

SECTION IV: OUT OF THE BLUE CREATION

1. Billy Collins, "The Blue," in *The Apple that Astonished Paris* (Fayetteville: University of Arkansas Press, 1988).

Chapter 19: How Much Her Do You Want, Henry?

1. W. Marslen-Wilson, "Biographical Interviews with H.M." (unpublished transcript; Cambridge, MA, MIT., 1970), pp. 1–144. Digitized and edited by Lori James and Don MacKay, for posting at http://mackay.bol.ucla.edu/ with permission from William Marslen-Wilson.

2. D. G. MacKay, L. W. Johnson, V. Fazel, and L. E. James, "Compensating for Language Deficits in Amnesia I: H.M.'s Spared Retrieval Categories," *Brain Sciences* 3, no. 1 (2013): 262–93.

3. E. H. M. Wiig and W. Secord, *Test of Language Competence: The Metaphor Subtest of the Expanded Edition* (New York: The Psychological Corporation, Harcourt, Brace, Jovanovich, 1988).

4. D. G. MacKay, L. E. James, C. B. Hadley, and K. A. Fogler, "Speech Errors of Amnesic H.M.: Unlike Everyday Slips-of-the-Tongue," *Cortex* 47 (2011): 377–408.

5. See N. Bolognini and T. Ro, "Transcranial Magnetic Stimulation: Disrupting Neural Activity to Alter and Assess Brain Function," *Journal of Neuroscience* 30, no. 29 (2010): 9647–50.

6. S. Corkin, *Permanent Present Tense: The Unforgettable Life of the Amnesic Patient, H. M.* (New York: Basic Books, 2013). Although Dr. Corkin suggested that Henry's language deficits reflected unsuspected damage to his left hemisphere, this hypothesis leaves unexplained his parallel deficits involving visual cognition, a right hemisphere function; see chapter 13.

7. Like Henry's *I want some her,* Yeats's *bee-loud glade* combines word categories in an inappropriate manner: the standard word order is not bee + loud + glade, but glade + loud + bees, as in *The glade is loud with bees.*

8. I. Viskontaz, *Brain Myths Exploded: Lessons from Neuroscience* (Chantilly, VA: Great Courses, 2017).

9. See D. G. MacKay and R. Goldstein, "Creativity, Comprehension, Conversation, and the Hippocampal Region: New Data and Theory," *AIMS Neuroscience* 3, no. 1 (2016): 105–42, DOI: 10.3934/Neuroscience.2016.1.105. Available at http://www.aimspress.com/journal/neuroscience.

10. D. G. MacKay and L. E. James, "Sequencing, Speech Production, and Selective Effects of Aging on Phonological and Morphological Speech Errors," *Psychology and Aging* 19 (2004): 93–110.

11. Steven Pinker, *How the Mind Works* (New York: W. W. Norton, 1997).

12. Ibid.

13. George Miller invented the more general concept known as chunking to explain why we can remember seven plus or minus two words each containing five letters as readily as seven plus or minus two random letters. I extend Miller's concept of chunking to the brain here in order to highlight its neural basis. Henry was unable to form new chunks.

Chapter 20: What's New, Henry?

1. B. Milner, "Memory and the Temporal Regions of the Brain," in *Biology of Memory*, ed. K. H. Pribram and D. E. Broadbent (New York: Academic Press, 1970).

2. Ibid.

3. Gollin fragmented figure retrieved from B. Kolk and I. Whishaw, *Fundamentals of Human Neuropsychology*, 4th ed. (New York: W. H. Freeman, 1996), p. 447.

4. Milner, "Memory and the Temporal Regions."

5. D. G. MacKay and L. W. Johnson, "Errors, Error Detection, Error Correction and Hippocampal-Region Damage: Data and Theories," *Neuropsychologia* (2013): 2633–50.

6. E. T. Rolls et al., "Responses of Single Neurons in the Hippocampus of the Macaque Related to Recognition Memory," *Experimental Brain Research* 93 (1993): 299–306; M. Kutas and S. A. Hillyard, "Event-Related Brain Potentials to Grammatical Errors and Semantic Anomalies," *Memory and Cognition* 11, no. 5 (1983): 539–50; R. T. Knight, "Contribution of Human Hippocampal Region to Novelty Detection," *Nature* 383 (1996): 256–59; K. S. Graham, M. D. Barense, and A. C. H. Lee, "Going Beyond LTM in the MTL: A Synthesis of Neuropsychological and Neuroimaging Findings on the Role of the Medial Temporal Lobe in Memory and Perception," *Neuropsychologia* 48 (2010): 831–53.

7. Knight, "Contribution of Human Hippocampal Region."

8. Ibid.

9. MacKay and Johnson, "Errors, Error Detection."

10. Oliver Sacks, "The Abyss: Music and Amnesia," *New Yorker*, September 24, 2007.

11. M. D. Lemonick, *The Perpetual Now: A Story of Amnesia, Memory, and Love* (New York: Doubleday, 2017).

12. MacKay and Johnson, "Errors, Error Detection."

13. W. J. M. Levelt, *Speaking: From Intention to Articulation* (Cambridge, MA: MIT Press, 1989), pp. 478–82.

14. MacKay and Johnson, "Errors, Error Detection."

Chapter 21: Uninvited Invaders of the Mind

1. Jeff Wheelwright, "This Old Brain," *Discover Magazine*, September 9, 2017, pp. 27–31.

2. W. Marslen-Wilson, "Biographical Interviews with H.M." (unpublished transcript; Cambridge, MA: MIT, 1970). Posted at and retrieved from http://mackay.bol.ucla.edu/ with permission from William Marslen-Wilson.

3. D. M. Burke, D. G. MacKay, J. S. Worthley, and E. Wade, "On the Tip of the Tongue: What Causes Word Finding Failures in Young and Older Adults?" *Journal of Memory and Language* 30 (1991): 542–79.

4. D. M. Burke, J. K. Locantore, A. A. Austin, and B. Chae, "Cherry Pit Primes Brad Pitt: Homophone Priming Effects on Young and Older Adults' Production of Proper Names," *Psychological Science* 15, no. 3 (2004): 164–70; see also L. E. James and D. M. Burke, "Phonological Priming Effects on Word Retrieval and Tip-of-the-Tongue Experiences in Young and Older Adults," *Journal of Experimental Psychology: Learning, Memory, and Cognition* 26 (2000): 1378–91.

5. Ibid.

6. James and Burke, "Phonological Priming Effects"; Burke, MacKay, Worthley, and Wade, "On the Tip of the Tongue."

7. Ibid.

8. S. Corkin, *Permanent Present Tense: The Unforgettable Life of the Amnesic Patient, H.M.* (New York: Basic Books, 2013).

9. B. J. Baars, J. Cohen, G. H. Bower, and J. W. Berry, "Some Caveats on Testing the Freudian Slip," in *Experimental Slips and Human Error: Exploring the Architecture of Volition*, ed. B. J. Baars (New York: Plenum, 1992), pp. 289–313.

10. D. G. MacKay, C. B. Hadley, and J. H. Schwartz, "Relations between Emotion, Illusory Word Perception, and Orthographic Repetition Blindness: Tests of Binding Theory," *Quarterly Journal of Experimental Psychology: Human Experimental Psychology*, 58A (2005): 1514–33.

11. Ibid.

12. D. G. MacKay, M. Shafto, J. K. Taylor, D. E. Marian, L. Abrams, and J. Dyer, "Relations between Emotion, Memory, and Attention: Evidence from Taboo Stroop, Lexical Decision, and Immediate Memory Tasks," *Memory & Cognition* 32 (2004): 474–88; D. G. MacKay and M. V. Ahmetzanov, "Emotion, Memory, and Attention in the Taboo Stroop Paradigm: An Experimental Analog of Flashbulb Memories," *Psychological Science* 16 (2005): 25–32; C. B. Hadley and D. G. MacKay, "Does Emotion Help or Hinder Immediate Memory? Arousal versus Priority-Binding Mechanisms," *Journal of Experimental Psychology: Learning, Memory, and Cognition* 32 (2006): 79–88.

13. D. G. MacKay, L. W. Johnson, and C. Hadley, "Compensating for Language Deficits in Amnesia II: H.M.'s Spared versus Impaired Encoding Categories," *Brain Sciences* 3, no. 2 (2013): 415–59; D. G. MacKay, L. W. Johnson, V. Fazel, and L. E. James, "Compensating for Language Deficits in Amnesia I: H.M.'s Spared Retrieval Categories," *Brain Sciences* 3, no. 1 (2013): 262–93.

14. Steven Pinker, *How the Mind Works* (New York: Norton, 1997).

15. Ibid.

16. G. Rizzolatti, "The Mirror-Neuron System and Imitation," in *Perspectives on Imitation: From Mirror Neurons to Memes*, ed. S. Hurley and N. Chater (Cambridge, MA: MIT Press, 2004); G. Rizzolatti et al., "Premotor Cortex and the Recognition of Motor Actions," *Cognitive Brain Research* 3, no. 131 (1996): 131–41.

17. M. M. Keane, J. D. E. Gabrieli, and S. Corkin, "Multiple Relations between Fact-Learning and Priming in Global Amnesia," *Society for Neuroscience Abstracts* 13 (1987): 1454. See also M. M. Keane, J. D. E. Gabrieli, H. C. Mapstone, et al., "Double Dissociation of Memory Capacities after Bilateral Occipital-Lobe or Medial Temporal-Lobe Lesions," *Brain* 118, no. 5 (October 1995): 1129–48, https://academic.oup.com/brain/article-abstract/118/5/1129/253968.

18. Sigmund Freud, *The Psychopathology of Everyday Life* (Berlin, 1901).

Chapter 22: How Creative Are You, Henry?

1. Steven Pinker, *How the Mind Works* (New York: Norton, 1997).

2. S. Corkin, *Permanent Present Tense: The Unforgettable Life of the Amnesic Patient, H.M.* (New York: Basic Books, 2013).

3. D. G. MacKay and R. Goldstein, "Creativity, Comprehension, Conversation, and the Hippocampal Region: New Data and Theory," *AIMS Neuroscience* 3, no. 1 (2016): 105–42.

4. W. Marslen-Wilson, "Biographical Interviews with H.M." (unpublished transcript; Cambridge, MA: MIT, 1970). Posted at and retrieved from http://mackay.bol.ucla.edu/ with permission from William Marslen-Wilson.

5. Corkin, *Permanent Present Tense.*

Chapter 23: You're Not Kidding, Henry

1. Richard W. Armour, *Punctured Poems: Famous First and Infamous Second Lines* (New York: New American Library, 1982).

2. S. Corkin, *Permanent Present Tense: The Unforgettable Life of the Amnesic Patient, H. M.* (New York: Basic Books, 2013), pp. xv, 17, 112–13, 121, 209, 234, 276, 284, 305, 307; B. Kolk and I. Whishaw, *Fundamentals of Human Neuropsychology*, 4th ed. (New York: W. H. Freeman, 1996), p. 447; B. Milner, S. Corkin, and H. L. Teuber, "Further Analysis of the Hippocampal Amnesic Syndrome: 14-Year Follow-Up Study of H. M." *Neuropsychologia* 6 (1968): 215–34; P. J. Hilts, *Memory's Ghost: The Strange Tale of Mr. M. and the Nature of Memory* (New York: Simon & Schuster, 1995), pp. 115–16.

3. D. G. MacKay, R. Stewart, and D. M. Burke, "H.M. Revisited: Relations between Language Comprehension, Memory, and the Hippocampal System," *Journal of Cognitive Neuroscience* 10, no. 3 (1998): 377–94; D. G. MacKay, L. E. James, and C. Hadley, "Amnesic H.M.'s Performance on the Language Competence Test: Parallel Deficits in Memory and Sentence Production," *Journal of Experimental and Clinical Neuropsychology* 30 (2008): 280–300.

4. For evidence on Henry's emotions and comprehension of ambiguity and metaphors, see: N. Hebben, S. Corkin, H. Eichenbaum, and K. Shedlack, "Diminished Ability to Interpret and Report Internal States after Bilateral Medial Temporal Resection: Case H.M.," *Behavioral Neuroscience* 99 (1985): 1031–39; MacKay, Stewart, and Burke, "H.M. Revisited"; MacKay, James, and Hadley, "Amnesic H.M.'s Performance on the Language Competence Test."

5. Joseph Farris, "The Pill! The Pill!," *Look*, December 30, 1963.

6. From W. Marslen-Wilson, "Biographical Interviews with H.M." (unpublished transcript; Cambridge, MA: MIT, 1970). Posted at and retrieved from http://mackay.bol.ucla.edu/ with permission from William Marslen-Wilson. Emphasis in the original. Dot strings indicate hesitations of corresponding lengths. Henry fabricated the idea that pills could make soap suds. Soap pods did not exist in 1970.

7. D. G. MacKay, L. W. Johnson, V. Fazel, and L. E. James, "Compensating for Language Deficits in Amnesia I: H.M.'s Spared Retrieval Categories," *Brain Sciences* 3, no. 1 (2013): 262–93.

8. Marslen-Wilson, "Biographical Interviews with H.M."

9. D. G. MacKay et al., "Speech Errors of Amnesic H. M.: Unlike Everyday Slips-of-the-Tongue," *Cortex* 47 (2011): 377–408.

10. E. L. Abel and M. L. Kruger, "Smile Intensity in Photographs Predicts Longevity," *Psychological Science*, 21 (2010): 542–44; L. S. Berk, S. A. Tan, W. F. Fry, et al., "Neuroendocrine

and Stress Hormone during Mirthful Laughter," *American Journal of the Medical Sciences* 298, no. 6 (1989): 390–96; Ed Diener and Micaela Y. Chan, "Happy People Live Longer: Subjective Well-Being Contributes to Health and Longevity," *Applied Psychology: Health and Well-Being* 3, no. 1 (2011): 1–43, doi:10.1111/j.1758-0854.2010.01045.x; R. T. Howell, M. L. Kern, and S. Lyubomirsky, "Health Benefits: Meta-Analytically Determining the Impact of Well-Being on Objective Health Outcomes," *Health Psychology Review* 1 (2007): 83–136; H. Nabi, M. Kivimaki, R. De Vogli, et al., "Positive and Negative Affect and Risk of Coronary Heart Disease: Whitehall II Prospective Cohort Study," *British Medical Journal* 337 (2008): 32–36; L. S. Richman, L. Kubzansky, J. Maselko, I. Kawachi, P. Choo, and M. Bauer, "Positive Emotion and Health: Going Beyond the Negative," *Health Psychology* 24 (2005): 422–29.

11. L. S. Berk, D. L. Felten, S. A. Tan, B. B. Bittman, and J. Westengard, "Modulation of Neuroimmune Parameters during the Eustress of Humor-Associated Mirthful Laughter," *Alternative Therapies in Health and Medicine* 7, no. 2 (2001): 62–72; T. Meyer, T. Smeets, T. Giesbrecht, C. W. E. M. Quaedflieg, and H. Merckelbach, "Acute Stress Differentially Affects Spatial Configuration Learning in High and Low Cortisol-Responding Healthy Adults," *European Journal of Psychotraumatology* 4 (2013): 74–76; D. J.-F. de Quervain, B. Roozendaal, and J. L. McGaugh, "Stress and Glucocorticoids Impair Retrieval of Long-Term Spatial Memory," *Nature* 394 (1998): 787–90.

12. Corkin, *Permanent Present Tense.*

13. P. Glenn and E. Holt, eds., *Studies of Laughter in Interaction* (London: Bloomsbury, 2013).

14. Corkin, *Permanent Present Tense.*

15. M. C. Duff, R. Gupta, J. A. Hengst, D. Tranel, and N. J. Cohen, "The Use of Definite References Signals Declarative Memory: Evidence from Patients with Hippocampal Amnesia," *Psychological Science* 22, no. 5 (2011): 666–73.

16. Corkin, *Permanent Present Tense.*

17. N. Hebben, S. Corkin, H. Eichenbaum, and K. Shedlack, "Diminished Ability to Interpret and Report Internal States after Bilateral Medial Temporal Resection: Case H.M.," *Behavioral Neuroscience* 99 (1985): 1031–39.

18. J. R. Lackner, "Observations on the Speech Processing Capabilities of an Amnesic Patient: Several Aspects of H.M.'s Language Function," *Neuropsychologia* 12 (1974): 199–207.

19. D. G. MacKay, R. Stewart, and D. M. Burke, "H.M. Revisited: Relations between Language Comprehension, Memory, and the Hippocampal System," *Journal of Cognitive Neuroscience* 10, no. 3 (1998): 377–94.

20. D. Hassabis, D. Kumaran, S. D. Vann, and E. A. Maguire, "Patients with Hippocampal Amnesia Cannot Imagine New Experiences," *Proceedings of the National Academy of Sciences* 104 (January 30, 2007): 1726–31.

21. Luke Dittrich, *Patient H.M.: A Story of Memory, Madness, and Family Secrets* (New York: Random House, 2016).

22. H. H. Brownell and H. Gardner, "Neurological Insights into Humor," in *Laughing Matters: A Serious Look at Humor*, ed. J. Durant and J. Miller (Essex: Longman Scientific and Technical, 1988), pp. 17–34.

23. From MacKay, James, Hadley, and Fogler, "Speech Errors of Amnesic H.M."; Robert Mankoff, "How about Never?" *New Yorker*, May 3, 1993.

24. Ibid.

25. Hollywood endings are especially effective completions for jokes. See Brownell and Gardner, "Neurological Insights into Humor."

26. MacKay, James, Hadley, and Fogler, "Speech Errors of Amnesic H.M."

27. A huge literature shows that positive attitudes are associated with good health, including: Nabi, Kivimaki, De Vogli, Marmot, et al., "Positive and Negative Affect and Risk of Coronary Heart Disease"; Richman, Kubzansky, Maselko, et al., "Positive Emotion and Health." For a review, see Diener and Chan, "Happy People Live Longer."

Chapter 24: Play It Again, Henry

1. W. Marslen-Wilson, "Biographical Interviews with H. M." (unpublished transcript; Cambridge, MA: MIT, 1970), pp. 1–144. Digitized and edited by Lori James and Don MacKay. Retrieved from http://mackay.bol.ucla.edu/.

2. D. G. MacKay, D. M. Burke, and R. Stewart, "H.M.'s Language Production Deficits: Implications for Relations between Memory, Semantic Binding, and the Hippocampal System," *Journal of Memory and Language* 38 (1998): 28–69.

3. P. J. Hilts, *Memory's Ghost: The Strange Tale of Mr. M. and the Nature of Memory* (New York: Simon & Schuster, 1995), pp. 115–16.

4. MacKay, Burke, and Stewart, "H.M.'s Language Production Deficits."

5. Ibid.

6. Ibid.

7. E. H. M. Wiig and W. Secord, *Test of Language Competence: The Metaphor Subtest of the Expanded Edition* (New York: Psychological Corporation, Harcourt, Brace, Jovanovich, 1988).

8. D. G. MacKay, L. W. Johnson, and C. Hadley, "Compensating for Language Deficits in Amnesia II: H.M.'s Spared versus Impaired Encoding Categories," *Brain Sciences* 3, no. 2 (2013): 415–59.

9. Ibid.

10. Ibid.

11. A. P. Shimamura, J. M. Berry, J. A. Mangels, et al., "Memory and Cognitive Abilities in University Professors: Evidence for Successful Aging," *Psychological Science* 6, no. 5 (1995): 271–77.

12. Ibid.

13. Ibid.

Chapter 25: Do You Remember What's-Her-Name, Henry?

1. K. H. McWeeny, A. W. Young, D. C. Hay, and A. W. Ellis, "Putting Names to Faces," *British Journal of Psychology* 78 (1987): 143–49; D. M. Burke, D. G. MacKay, J. S. Worthley, and E. Wade, "On the Tip of the Tongue: What Causes Word Finding Failures in Young and Older Adults?" *Journal of Memory and Language* 30 (1991): 542–79; L. E. James, "Meeting Mr. Farmer versus Meeting a Farmer: Specific Effects of Aging on Learning Proper Names," *Psychology and Aging* 19, no. 3 (2004): 515–22.

2. N. J. Slamecka and P. Graf, "The Generation Effect: Delineation of a Phenomenon," *Journal of Experimental Psychology* 4, no. 6 (1978): 592–604.

3. T. K. Landauer and R. A. Bjork, "Optimum Rehearsal Patterns and Name Learning," in *Practical Aspects of Memory*, ed. M. Gruneberg, P. E. Morris, and R. N. Sykes (London: Academic Press), pp. 625–32.

4. W. Marslen-Wilson, "Biographical Interviews with H.M." (unpublished transcript; Cambridge, MA: MIT, 1970). Posted at and retrieved from http://www.mackay.bol.ucla.edu/ with permission from William Marslen-Wilson.

5. Ibid. See also S. Corkin, *Permanent Present Tense: The Unforgettable Life of the Amnesic Patient, H.M.* (New York: Basic Books, 2013).

6. Marslen-Wilson, "Biographical Interviews with H.M."

7. Ibid.

8. Ibid.

9. D. G. MacKay, L. W. Johnson, and C. Hadley, "Compensating for Language Deficits in Amnesia II: H.M.'s Spared versus Impaired Encoding Categories," *Brain Sciences* 3, no. 2 (2013): 415–59; D. G. MacKay, L. W. Johnson, V. Fazel, and L. E. James, "Compensating for Language Deficits in Amnesia I: H.M.'s Spared Retrieval Categories," *Brain Sciences* 3, no. 1 (2013): 262–93.

10. Like Henry, "Mickey"—an amnesic patient examined by Harvard University Professor Dan Schacter—could easily learn newly encountered facts about *proper names*. Despite being unable to learn other types of new facts, Mickey quickly learned that *Hoboken* hosted the first baseball game and that *Theodore Roosevelt* held the world record for number of hands shaken. (See D. Schacter, *The Seven Sins of Memory: How the Mind Forgets and Remembers* [New York: Houghton Mifflin, 2001], pp. 165–66).

11. MacKay, Johnson, and Hadley, "Compensating for Language Deficits in Amnesia II."

12. Ibid.

13. Benefits aside, Henry's invented proper name often baffled his listeners. They needed a proper introduction such as *Let's call this man David*. Henry skipped this basic step for a reason: To say *Let's call her X or Let's call this man Y*, Henry would have had to use the very categories he was trying to avoid—pronouns and phrases containing common nouns.

14. MacKay, Johnson, and Hadley, "Compensating for Language Deficits in Amnesia II."

15. Henry's responses appeared in Marslen-Wilson, "Biographical Interviews with H.M."

16. MacKay, Johnson, and Hadley, "Compensating for Language Deficits in Amnesia II; MacKay, Johnson, Fazel, and James, "Compensating for Language Deficits in Amnesia I: H.M.'s Spared Retrieval Categories," *Brain Sciences* 3, no. 1 (2013): 262–93.

17. From Marslen-Wilson, "Biographical Interviews with H.M."

18. Corkin, *Permanent Present Tense*.

19. MacKay, Johnson, and Hadley, "Compensating for Language Deficits in Amnesia II."

Chapter 26: Let's Celebrate a Promise Fulfilled, Henry

1. S. Corkin, *Permanent Present Tense: The Unforgettable Life of the Amnesic Patient, H. M.* (New York: Basic Books, 2013), p. 287.

2. Ibid., pp. 295–96.

3. Ibid., p. 287.

4. Daniel Kahneman, *Thinking, Fast and Slow* (New York: Farrar, Straus & Giroux, 2011), pp. 1–499.

5. D. G. MacKay and L. E. James, "Visual Cognition in Amnesic H.M.: Selective Deficits on the What's-Wrong-Here and Hidden-Figure Tasks," *Journal of Experimental and Clinical Neuropsychology* 31 (2009): 769–89.

6. A. C. Lee, M. D. Barense, and K. S. Graham, "The Contribution of the Human Medial Temporal Lobe to Perception: Bridging the Gap between Animal and Human Studies," *Quarterly Journal of Experimental Psychology* 58B (2005): 300–25; A. C. Lee, T. J., Bussey, E. A. Murray, et al., "Perceptual Deficits in Amnesia: Challenging the Medial Temporal Lobe 'Mnemonic' View," *Neuropsychologia* 43 (2005): 1–11; M. D. Barense, D. Gaffan, and K. S. Graham, "The Human Medial Temporal Lobe Processes Online Representations of Complex Objects," *Neuropsychologia* 45 (2007): 2963–74; M. D. Barense, T. J. Bussey, A. C. Lee, et al., "Functional Specialization in the Human Medial Temporal Lobe," *Journal of Neuroscience* 25 (2005): 10239–46.

7. D. G. MacKay, L. W. Johnson, and C. Hadley, "Compensating for Language Deficits in Amnesia II: H.M.'s Spared versus Impaired Encoding Categories," *Brain Sciences* 3, no. 2 (2013): 415–59; D. G. MacKay, L. W. Johnson, V. Fazel, and L. E. James, "Compensating for Language Deficits in Amnesia I: H.M.'s Spared Retrieval Categories," *Brain Sciences* 3, no. 1 (2013): 262–93.

8. Sarah E. MacPherson and Sergio Della Sala, *Single Case Studies of Memory* (in press); D. G. MacKay, "The Earthquake That Reshaped the Intellectual Landscape of Memory, Mind and Brain: Case H.M.," in *Single Case Studies of Memory*, ed. S. E. MacPherson and S. D. Sala (in press).

9. D. Z. Hambrick, T. A. Salthouse, and E. J. Meinz, "Predictors of Crossword Puzzle Proficiency and Moderators of Age-Cognition Relations," *Journal of Experimental Psychology: General* 128 (1999): 131–64.

INDEX

Pages in *italics* indicate figures.